Praise for *AI for Social Good*

An inspiring overview of what machine learning and artificial intelligence can already do to make the world better, and what can be done to use these tools more effectively.

—Andrew Gelman
Professor of Statistics and Political Science,
Columbia University

Our evolution has been an ascent toward increasing consciousness, with tools, fire, domestication of animals, and agriculture as stepping-stones along the way. Computers, the internet, and now AI have emerged rapidly to become a major part of the technological landscape. Rahul Dodhia's book *AI for Social Good* is a comprehensive exploration of this transformative field, which for technologically challenged non-experts like myself, brilliantly demystifies this exciting field. It leaves you with the hope that AI will be harnessed for the good of the planet and used ethically and responsibly. This book is a must-read for anyone interested in understanding AI's past and present, as well as its profound influence on the future of humanity on planet Earth.

—Louise Leakey
Paleoanthropologist,
Turkana Basin Institute, Kenya

In his important new book, leading AI practitioner Rahul Dodhia takes us on a highly accessible whirlwind tour of how AI works, what it can and cannot do, and why it sometimes goes off the rails. You will find inspirational stories on how AI can be used for good and cautionary tales that temper your hubris. Bringing his considerable experience to bear, Dodhia's compelling nuts-and-bolts discussion of how to set up teams that make the most of this potentially transformative technology is a must-read for anyone leading AI-based projects. This book has something in it for everyone seeking to understand and make the most of this rapidly evolving tool.

—Jacob N. Shapiro
Director, Empirical Studies of Conflict Project,
Princeton University

Just as the dawn of the nuclear age simultaneously shaped our hopes and our greatest fears for the future of the planet in the last century, so too for artificial intelligence in our century. The world's deeply vulnerable environment and its communities are at a crossroads: one path leading to ecosystem collapse, triggering extreme poverty and violence, the other toward balance and recovery. Our ability and determination to choose the right path are profoundly linked to the choices we make on the use of this nascent technology. In his writing, Rahul provides the critical thinking to channel our choices on the use of AI into a force for long overdue positive change.

—Emmanuel de Merode
Director of Virunga National Park

AI for Social Good is a compelling book that explores the responsible use of AI as a force for positive and transformative change. It offers a valuable guide for those interested in leveraging AI to tackle the urgent challenges of our time. Against the backdrop of our rapidly changing world and the unprecedented threats we face, the book provides concrete examples of how AI is already playing a crucial role in enhancing our understanding of, preparedness for, and response to global challenges. These examples range from the development of early warning systems for droughts and rapid disaster response programs to aiding decision-making in support of food and water security and the creation of innovative medical solutions.

Rahul, a prominent voice in the emerging AI for Social Good movement, underscores the potential of this rapidly advancing technology as an "essential ingredient in our efforts to create a better world for future generations."

His book offers an insightful overview of AI and its evolution, providing tangible examples of its diverse applications and ability to drive positive change. It underscores the critical importance of ethics and regulation in the field of AI and provides a glimpse into the future technologies that will further propel its applications and impact.

In an era where AI is both celebrated and met with significant apprehension, *AI for Social Good* serves as an excellent guide, offering practical advice, real-world examples, and a compelling vision for harnessing AI to address our most pressing challenges. It is an invaluable resource for anyone navigating the rapidly evolving AI landscape in the pursuit of societal betterment. It serves as a resounding call to action, encouraging individuals to become part of the movement for positive change.

—Inbal Becker-Reshef
Program Director of NASA Harvest

AI FOR SOCIAL GOOD

RAHUL DODHIA

AI FOR SOCIAL GOOD

USING ARTIFICIAL INTELLIGENCE TO SAVE THE WORLD

WILEY

Published by John Wiley & Sons, Inc., Hoboken, New Jersey.
Published simultaneously in Canada.

For general information on our other products and services or for technical support, please contact our Customer Care Department within the United States at (800) 762-2974, outside the United States at (317) 572-3993 or fax (317) 572-4002.

Wiley also publishes its books in a variety of electronic formats. Some content that appears in print may not be available in electronic formats. For more information about Wiley products, visit our web site at www.wiley.com.

Library of Congress Cataloging-in-Publication Data:

Names: Dodhia, Rahul, author.
Title: AI for social good / Rahul Dodhia.
Description: Hoboken, New Jersey : Wiley, [2024] | Includes index.
Identifiers: LCCN 2023047408 (print) | LCCN 2023047409 (ebook) | ISBN 9781394205783 (cloth) | ISBN 9781394205837 (adobe pdf) | ISBN 9781394205790 (epub)
Subjects: LCSH: Artificial intelligence—Moral and ethical aspects.
Classification: LCC Q334.7 .D634 2024 (print) | LCC Q334.7 (ebook) | DDC 174/.90063—dc23/eng/20231103
LC record available at https://lccn.loc.gov/2023047408
LC ebook record available at https://lccn.loc.gov/2023047409

Cover Design: C. Wallace
Cover Image: © EpicEtch / Adobe Stock
Author Photo: Courtesy of the Author

SKY10064821_011524

Dedicated to my late mother, whose memory still guides me, and my father, who taught me compassion and to care for the world.

Contents

Acknowledgments

MANY OF THE examples in the book come from projects led by past and present members of the AI for Good Research Lab. Their intelligence and dedication to improving the world around them inspire me every day. For the work they have done, thanks to Anthony Ortiz, Zhongqi Miao, Caleb Robinson, Meghana Kshirsagar, Simone Fobi Nsutezo, Juan Lavista Ferres, Shahrzad Gholami, Felipe Oviedo, Thomas Roca, Akram Zaytar, Gilles Hacheme, Lucas Meyer, Girmaw Abebe Tadesse, Md Nasir, Mayana Pereira, Yixi Xu, Darren Tanner, Amrita Gupta, Will Fein, Tammy Glazer, Anusua Trivedi, Siyu Yang, Ming Zhong, Hyojin Song, Sumit Mukherjee, and John Kahan.

Thanks also to Dan Morris, who personifies the AI for Good ethos, and Cameron Birge, who helped bring about several of the projects discussed in the book. And special thanks to the larger AI for Good Lab at Microsoft, of which the Research Lab is a part, for continually fueling the momentum of the AI for Social Good movement.

Finally, thanks are also due to my daughter, Arya, for her inquisitive nature that keeps me on my toes and her frequent wrestling matches, providing much-needed screen breaks. I also owe gratitude to my wife, Annette de Soto, who reviewed this book and offered questions, revisions, and a close reading of the text honed from too many years at the University of Chicago.

About the Author

Rahul Dodhia heads the AI for Good Research Lab at Microsoft, based in Redmond, Washington. He leads a team of AI researchers dedicated to addressing global challenges using artificial intelligence. His work focuses on sustainability, humanitarian action, and health issues, paying special attention to climate adaptation in the Global South.

Prior to his current role, he led machine learning teams at several corporations, including eBay, Amazon, and Expedia. He also served at the NASA Ames Research Center, where he applied foundational research on human memory to address safety concerns in general aviation and space flight.

Rahul's undergraduate education was at Brandeis University, earning a BA in Mathematics, with the highest honors, summa cum laude. His journey into the world of artificial intelligence began during his graduate studies in the psychology department at Columbia University. He conducted extensive research on human memory and decision-making models there, ultimately earning his PhD.

Rahul grew up in Thika, Kenya, a place that has seen profound ecological change. In addition to his research interests, he was a competitive sheepherder with his beloved Border collie, Artoo Dogtoo.

Introduction

In 2022, the world was horrified by the earthquake that devastated Turkey and Syria. Like many people around the world, my team at Microsoft, the AI for Good Research Lab, wondered how we could help from so far away. Having previously utilized satellite imagery to identify areas of destruction, the Lab sprang into action, providing maps of areas in need to the authorities. When the historic town of Lahaina in Hawaii was engulfed in flames the following year, we supported the American Red Cross with maps with localized estimates of destruction, enabling them to disburse aid in record time to those most in need. Meanwhile, in drought and locust-stricken Kenya, we collaborated with the Nature Conservancy to identify smallholder farms and devise irrigation solutions. In the United States, as disinformation endangered lives and democracy, we developed tools to assess and trace the origins of false information. These initiatives all had in common new computing tools developed within the last few years: artificial intelligence that mimicked the neuronal processes of living brains.

At Microsoft's AI for Good Research Lab, my team dedicates itself daily to tackling humanity's global challenges using artificial intelligence. Despite numerous instances of AI being employed for positive purposes, many remain unaware of this side of the story of AI. Inspired by the work of the Lab, I wrote *AI for Social Good* for those looking to grasp the basics of AI and its real-world applications that affect positive change in society. The book clarifies AI concepts and offers a lucid and direct explanation of the technology and its numerous applications for positive impact. Whether you are new to the AI world or already working with AI, I hope this book will enhance your understanding and spark innovative applications of AI for the greater good.

Interest in artificial intelligence surged in 2023, catalyzed by the remarkable launch of ChatGPT. For generations that grew up with narratives of robots and computers with human-like intelligence, it appeared as if the future had finally arrived. However, admiration for large language models like ChatGPT has been dampened by their inclination to lead people astray. Concerns about AI's rapid, unchecked development have become louder, and respected AI researchers and leaders in technology have joined in with warnings that technology is advancing at a pace greater than our ability to absorb it. The speed at which AI is evolving makes it difficult to accurately predict its outcomes, underscoring the urgent need for a comprehensive set of guidelines to navigate this uncharted territory. Many of us are now advocating for the incorporation of ethical principles at the heart of AI development.

All of this is unfolding against a backdrop of significant transformation in the global ecosystem. Beyond perennial issues, such as employment and livelihoods, exacerbated by fears of AI usurping them, we are now also confronted with the challenges posed by climate change. Natural disasters may be growing more devastating, and food and water insecurity are rearing their ugly heads. This multitude of problems seemed overwhelming, but now AI offers some hope. We may be on the verge of discovering new solutions to these problems.

A movement that can be termed AI for Social Good has arisen to counter the dystopian narrative of AI that builds on fears of economic setbacks and global war. It manifests in various ways, from nonprofit organizations to private sector projects, from academic conferences to online communities. It is not an organized movement where members pay dues and have newsletters. But it captures the spirit of people who are troubled by what they see coming in the future, and it has been embraced by dedicated young people with a burning desire to be a part of the solution.

The book is structured to be read sequentially, but each chapter stands on its own so readers with particular interests can jump around. Here is a brief summary of each chapter.

Chapter 1 traces a brief history of artificial intelligence, how it arose from the early days of computing in the 19th century to its emergence, in fits and starts, within the last few decades. This foundation for understanding AI's development highlights key individual achievements while acknowledging the collective efforts of their peers.

Chapter 2 is a textbook-style exposition of the components that constitute AI. It introduces the reader to the terms commonly used by practitioners of AI. Terms such as neural networks, machine learning, and large language models are explained here. The history of AI from the previous chapter is appended by more stories of how technical aspects of AI came into being.

Chapter 3 highlights AI's potential to drive positive change for the reader to envision novel ways in which AI can be harnessed to address the pressing issues of our time. Several examples of how AI is used for social good are given, with an emphasis on humanitarian and environmental issues. It explores how newly available data, such as satellite and drone images and recent advancements like foundation models for language, creates opportunities for breakthroughs in the challenges plaguing society.

Chapter 4 continues the discussion of AI for social good but focuses more on scientific endeavors. By showcasing the real-world applications and implications of AI in these crucial scientific domains, the chapter aims to enlighten the reader on the indispensable role of AI in addressing contemporary scientific challenges and advancing human knowledge. Examples from biodiversity, astronomy, and proteomics illustrate this impact. The reader is not expected to have prior knowledge of these fields, and Chapter 4 introduces their significance.

Chapter 5 dispels the notion of an AI utopia. It addresses the potential pitfalls of AI and explains the fears raised by prominent technologists, again with several examples. We look particularly at how AI can supercharge propaganda and disinformation and how societal biases are mirrored in AI, a reflection of our own inclinations and actions. The chapter aims to foster a more nuanced understanding of the potential repercussions of AI, urging the reader to approach its development and deployment with a balanced perspective and a critical eye.

Chapter 6 elaborates on one of the book's central themes: AI development should be based on a core of ethics and agreed-upon standards. The need for regulation is necessary to mitigate the negative implications of AI. History shows us the need for reining in the more negative aspects of humanity, a sort of societal superego to balance the Id's baser instincts. This chapter explores the nuances of regulating AI by examining case studies and global approaches. It calls for international collaboration to establish guidelines protecting individual rights while allowing controlled experimentation. Core themes include transparency, consent, data security, algorithmic fairness, and human oversight for high-stakes decisions. Though an imperfect process, mindful governance of AI via laws, industry standards, and social norms is vital to realizing its benefits without unacceptable risks.

In **Chapter 7**, I draw on my experience running AI teams to offer practical advice for constructing effective teams, bridging knowledge gaps, and aligning technical capabilities with real-world utility. Developing impactful AI requires a team with diverse expertise, effective collaboration, and core roles like the project manager, domain expert, and AI expert who each contribute unique perspectives. Frequent communication and feedback loops ensure the AI model matches real-world requirements. However, challenges inevitably arise regarding data quality, model accuracy, and ethical implications. A thoughtful, human-centric approach is crucial, with human oversight playing a pivotal role in deploying reliable AI.

Chapter 8, the last chapter, looks ahead to future technologies and the immense changes that AI might wreak on our society. We are merely at the beginning of our journey with a new form of intelligence, with technologies already in the pipeline, such as quantum computing and DNA storage, that could radically redefine our conception of what we think it means to be human.

This book aspires to disseminate innovative ideas and serve as a source of inspiration for those eager to harness the power of AI to address some of the most critical challenges facing society today. If this book leaves you eager for more, an upcoming book going deeper into the topics covered here will be coming soon. Authored by several members of the AI for Good Research Lab, it will be a non-technical but in-depth discussion of the projects the lab has undertaken.

1

A Brief History of Artificial Intelligence

"Artificial intelligence is growing up fast, as are robots whose facial expressions can elicit empathy and make your mirror neurons quiver."

– Diane Ackerman

"The science of today is the technology of tomorrow"

– Edward Teller

IN 1997, IBM's Deep Blue computer famously defeated world chess champion Garry Kasparov in a six-game match. This event marked a major milestone in the development of AI, as it demonstrated that a machine could outthink a human in a complex game with countless possible moves. The jubilation felt on achieving such a feat was mixed with hand-wringing that the age of machines was about to eclipse the age of humankind. Kasparov himself could not believe a machine could have defeated him and insisted this was a modern version of the Mechanical Turk, a 19th century con where a small person hid inside a supposed automaton and played chess.[1,2] Despite these expressions of disbelief, the match captured the world's attention. Chess was, after all, an ancient game highly revered as an expression of human mental ability. This event sparked a new interest in the abilities of machines that could think and adapt and even outshine humans.

Nearly seven decades since the prefix "artificial" was attached to intelligence, we live on the cusp of one of the largest disruptions in human society. When the CEO of Google, Sundar Pichai, calls AI one of humanity's most profound inventions,[3] and other tech luminaries such as Bill Gates argue, "The development of AI is as fundamental as the creation of the microprocessor, the personal computer, the Internet, and the mobile phone,"[4] and Elon Musk goes so far as to deem it potentially more dangerous than nuclear weapons,[5] it is hard to dismiss the furor around this new technology as hyperbole. We may indeed be living in a time of profound change.

Artificial intelligence's rise and awesome potential have been a topic of discussion among tech insiders for quite some time. Now, with the emergence of ChatGPT, a much greater slice of humanity is witnessing firsthand the impact of this technology in their daily lives. If there are skeptics questioning the impact and abilities of artificial intelligence, their doubts are certainly being challenged.

AI manifests in our lives in the form of self-driving cars, virtual assistants such as Alexa and Siri, and unprecedented information via search engines. It is even more prevalent behind the scenes, powering medical assistants, farming, and disaster response. AI developments are quickly transforming the way we work, communicate, and even think. The invention of the automobile changed landscapes and economies, while radio and telephone transformed communications and society. AI is poised to join these ranks of major disruptors in the coming years. We are witnessing the birth of a transformative force that will change how we make decisions and perceive the world around us.

However, the implications of this technological transformation are not without their challenges. There are concerns over privacy, security, and job displacement. Evidence shows that AI reflects some of society's worst habits, such as racial and societal bias. As AI continues to become more sophisticated and more integral to our lives, individuals and society must carefully consider its ethical implications. With the proper safeguards in place, the undeniable benefits of AI could usher in a new era of progress and prosperity for all.

How Innovators Throughout History Paved the Way for Modern AI: From Babbage to Turing

Artificial intelligence was long the province of fiction, fantasy, folklore, and myth. Inanimate objects developing human-like intelligence and abilities beyond our own are common in the stories we share. From figures such as mystical golems in Jewish tales and enigmatic homunculi of the Middle Ages to the evil computer HAL in *2001: A Space Odyssey* and the iconic droids in *Star Wars*, these legends reflect our curiosity and desire to create intelligence in our image.

Next, we trace the broad outlines of AI's emergence, from early conceptualizations of universal calculating machines to the first manifestations of what we today call AI.

Charles Babbage

The first practical steps toward AI happened in the last 200 years. Charles Babbage (Figure 1.1) emerged as a seminal figure in the history of AI, revered by many as the progenitor of this field. Babbage, a brilliant mathematician and inventor, possessed an indomitable spirit, a penchant for spectacle, and an insatiable curiosity that led him to his brilliant achievements in computing.[6,7] His fascination with automatons mimicking human intelligence was sparked at age eight when his mother whisked

Figure 1.1 Drawing of Charles Babbage

Credit: The Illustrated London News / Wikimedia Commans / Public Domain.

him away to a museum of scientific artifacts and wonders. There, he saw an artful creation—a dancer cradling a bird—so exquisitely crafted that it appeared lifelike. From that moment forward, Babbage's destiny was irrevocably entwined with the pursuit of crafting machines capable of emulating human behavior.

In his late 20s in the early 1800s, Babbage designed the first mechanical computer, the Difference Engine. This groundbreaking machine could perform complex mathematical calculations, such as producing tables of logarithms.[8,9] Indulging his showman tendencies, Babbage delighted in donning extravagant attire as he showcased his creation to the venerable Royal Society in London and other esteemed venues across England.

Tales of his eccentricities, ranging from chasing musicians away from his abode when they impinged on his concentration to his fastidious craftsmanship, where gears and tools personally ground by him remained in use long after his death, embellished the legend of this extraordinary man.

The Difference Engine was never completed during Babbage's lifetime. It wasn't until the 1990s that it was finally built according to Babbage's design. It is on display at the London Science Museum, and a second one remains in the possession of a private donor who financed its creation.

Although Babbage was not able to see his design take life, it inspired his later, more audacious creation, the Analytical Engine. This was a much more ambitious endeavor, surpassing the Difference Engine in its versatility. Babbage intended it to be a general-purpose computing machine that could be instructed to perform any type of calculation. He envisioned tables of mathematical values being formulated, and these tables of values would inform calculations of things like dates of eclipses. Crucially, the Analytical Engine encompassed the fundamental duality of modern computers: the ability to store and process vast troves of data.

Regrettably, quarrels with his engineers and the drying up of funding meant that the Analytical Engine, like the earlier Difference Engine, was never built. It nevertheless stands as a major milestone in the history of computing. It was the first machine designed to be truly programmable. And it also helped to popularize the idea of artificial intelligence.

Ada Lovelace

Now recognized as the world's first computer programmer, Ada Lovelace (Figure 1.2) collaborated with Charles Babbage on his prototypes. When recounting the history of science and technology, the contributions of women have often been overlooked or underrepresented. But Ada Lovelace, daughter of the romantic poet Lord Byron and Anne Isabelle Milbanke, left her mark as indelibly as any male pioneer. Despite being born in the 19th century, when women's opportunities were limited, Ada Lovelace defied societal norms and fervently pursued her passion for mathematics and science. Her mother was responsible in large part for Ada's education. Seeking to shelter Ada from her father's perceived and infamous instabilities, she ensured Ada got a firm grounding in logic and mathematics.[10]

Figure 1.2 Ada Lovelace, watercolor painting, possibly by Alfred Edward Chalon in 1840

Credit: Science Museum Group / Wikimedia Commans / Public Domain.

When she was 17, Ada Lovelace met Charles Babbage at the house of Mary Sommerfield, a Scottish scientist and mathematician. Sommerfield had recognized a keen scientific intelligence in Lovelace and consciously brought about this intellectual match. Lovelace and Babbage became collaborators.

Her insight into Babbage's Analytical Engine went beyond his own ideas. She envisioned its potential beyond mere calculation. She recognized that the Analytical Engine could be used for more than just crunching numbers; it could be a tool for creativity and generating complex outputs. Her notes included an algorithm for calculating Bernoulli numbers, which is widely regarded as the world's first computer program. This visionary insight earned her the title of the world's first computer programmer.

Unfortunately, like many bright intelligences, she succumbed to her body's infirmities at age 36. But her legacy in computer science guides researchers and engineers to this day.[11]

John von Neumann

John von Neumann (Figure 1.3) is another of the most prominent people to lay the foundations of computer science. Hailing from Budapest, Hungary, von Neumann was a child prodigy, a versatile intellectual who hungered for mathematics and physics. His unconventional, multidisciplinary approach to studying made many skeptical of his seriousness, and, like his predecessor Charles Babbage, he gained a reputation as a maverick.[12]

Figure 1.3 John von Neumann

Credit: Los Alamos National Laboratory / Wikimedia Commans / Public Domain.

Von Neumann's extraordinary intellect carried him to doctorates in chemical engineering at the University of Zurich and mathematics from the University of Prague. When he submitted his doctoral dissertation to the faculty at the University of Zurich, the professors found it so profound and complex that they couldn't fully understand it. They asked him to simplify it, but with characteristic conviction, he firmly declined. To his mind, if they failed to comprehend the magnitude of his ideas, they lacked the qualification to pass judgment upon them. As a result, his dissertation remained unfinished and was never formally submitted, yet it still significantly impacted the field of mathematics and was later published as a monograph.[13]

Von Neumann moved on to the University of Berlin, where he continued to baffle his peers and students. Many stories of his time there illustrate his brilliance. Once, a student in a statistics lecture asked him a challenging question about a complex mathematical calculation. Without skipping a beat, von Neumann proceeded to solve the problem mentally and provided the answer within seconds. His lectures were often marked by brilliant expositions, which the students would then spend hours deciphering amongst themselves.

In the 1930s, he landed a teaching appointment at Princeton University. There, his genius would shine most brilliantly, and his pioneering contributions would forever transform the field of computing. Today, we take for granted the CPU as the brain of a computer and memory where computer programs are stored. Von Neumann was the genius who formulated these concepts and helped make them a reality, like UNIVAC, one of the first computers ever built.[14]

Alan Turing

For decades, the Turing test was held up as the holy grail of computing and artificial intelligence. It was an answer to the question of how we would know when machines had become intelligent. Mathematician Alan Turing (Figure 1.4) proposed his eponymous test, though he called it the Imitation Game.[15,16] The test consists of questions posed to the machine and humans. If the answers are indistinguishable, one cannot tell which answers came from the machine, then the machine has won the game and passed the test.

Figure 1.4 Alan Turing

Credit: Dunk/Flickr/Public domain.

Until the early 2000s, beating the test seemed like a very difficult, nearly impossible task. This seemingly insurmountable challenge for artificial intelligence researchers imbued the Turing test with an aura of mystery and intrigue. It became a symbol of the quest for artificial intelligence.

The Turing test had profound philosophical implications. If a machine is equivalent to a human, then what does it say about human intelligence? What can it tell us about consciousness? It had practical implications as well, which we're now seeing firsthand. ChatGPT and DALL-E by OpenAI have taken the world by storm, and there's no doubt that ChatGPT can pass the Turing test.

The Turing test did, and still does, have its skeptics, who saw it as a limited indicator of machine intelligence.

They argued that relying on mimicking human speech patterns did not reflect on general intelligence. Now that we have reached an honest reckoning, it's unclear whether this holy grail is as significant as we thought it was. ChatGPT is undoubtedly very human-like in its responses, but it is clearly still a non-conscious machine.

The eponym of this test, Alan Turing, was an Englishman who led the successful effort to break the German code during World War II, and then developed his theories of computing at the National Physical Laboratory. While Babbage's work was foundational for computing, and von Neumann influenced architected computer designs, Turing was a pioneer of theoretical computer science and artificial intelligence. His notion was of a Universal machine, known as a Turing machine, that could compute anything given a set of instructions. If this sounds like Babbage's Analytical Engine, it's because fundamentally they both had the same underlying idea of a flexible computing machine. Turing's mathematical concept, though, laid practical foundations for the development of computers.

The history of artificial intelligence is populated by thousands of mathematicians, engineers, psychologists, and scientists. However, among this vast sea of contributors, these four pioneers serve as human faces for the early development of artificial intelligence.

The Emergence of Modern AI

From the 1950s onward, the story of AI has taken on a certain canonical shape, which will be sketched here. Like its older, more venerable cousin of theoretical physics, its coming to maturity is evolving into a story we tell ourselves and each other, a narrative that shapes our collective understanding. The story begins with Babbage and Lovelace and continues with Turing and von Neumann, and then comes one of the nodes, a turning point, in the 1950s.

The Dartmouth Conference: A Turning Point

In the summer of 1956, a group of researchers gathered on the campus of Dartmouth College to discuss a new field whose name had just been made up by one of the organizers. John McCarthy put "artificial intelligence" in the name of the conference and in the proposal for its funding.[17] The Dartmouth Conference was a gathering of some of the leading researchers in computer science, mathematics, philosophy, and psychology, and would come to be seen as AI's genesis moment.

The organizers were old friends. John McCarthy and Marvin Minsky had been roommates at Princeton University and had remained close friends ever since. Nathaniel Rochester and Claude Shannon were former colleagues from Bell Labs and collaborated on the development of computer languages and hardware. By most accounts, the gathering was somewhat chaotic, with a loose flow of ideas, brilliant minds each pursuing their own agendas, and people coming and going as they wished. Marvin Minsky brought his electric guitar and played late into the night, entertaining his colleagues with his musical skills.

It seemed nothing would come out of this gathering. Turing and von Neumann, who would have been expected to be major figures at the conference, were dead (Turing) or ailing (von Neumann).[18,19] But, in the years that followed the Dartmouth Conference, many of the participants went on to become leaders in the field of AI. John McCarthy, for instance, went on to develop the Lisp programming language, which became a vital tool in AI research. Marvin Minsky co-founded the MIT Artificial Intelligence Laboratory and became one of the most influential figures in the field. Nathaniel Rochester continued to work at IBM, overseeing the development of some of the earliest computer systems. Two of them, Nash and Simon, went on to win Nobel prizes for other endeavors.

Minsky continued the work he had begun during his doctoral research and shaped the research direction for the new field with his colleague Samuel Papert. Psychologists had been interested in how the brain worked and tried to model the behavior of individual cells. Frank Rosenblatt put together what would become the most famous neural network of all, the perceptron.[20] Incredibly simple compared to the monumental edifices that AI scientists now build, it was nevertheless an astounding demonstration of how cells could exhibit behavior. Minsky wrote his doctoral thesis on neural networks, and his book with Samuel Papert, *Perceptrons*, made the titular neural network famous.[21] Rather than celebrating Rosenblatt's perceptron, the book argues that the network was too simple. Consisting of just a single layer of artificial neurons, they could be used to solve only very basic problems. They proposed a theoretical multi-perceptron, a neural network with multiple layers that could handle more sophisticated tasks. Ironically, the harsh critique contained in the book about the perceptron, along with no practical way of implementing a multi-perceptron, was an early, unintended salvo that crashed the enthusiasm for AI.

From Optimism to Pessimism: The Story of the AI Winter

The 1960s were characterized by optimism and a focus on fundamental research. In comparison, the 1970s were a more challenging time for AI research, with a shift toward applied research and the development of expert systems. The enthusiasm of the 1950s and 1960s was exemplified by statements that promised human-level intelligence within a few years. *Life* magazine published this quote from an interview with Marvin Minsky, "[In] three to eight years, we will have a machine with the general intelligence of an average human being."[22] A few years later, in 1973, economist Herbert Simon, one of the creators of the world's first artificial intelligence program, famously declared that "machines will be capable, within twenty years, of doing any work a man can do."[23]

These predictions turned out to be wildly optimistic. The power of computers at the time wasn't enough to make their dreams a reality. Sure, computers had come a long way since the days of Turing machines that cracked the German codes in World War II, but the theory quickly outpaced the hardware. It was like trying to build a skyscraper with just hammers and nails.

The pendulum swung toward pessimism. In 1975, mathematician James Lighthill published a report for the British government that criticized the state of AI research at the time, arguing that progress had been "grossly exaggerated" and that the field was unlikely to deliver significant results in the near future.[24] He argued that the combinatorial explosion of choices that most decision processes would face would never be overcome. A few years earlier, philosopher Hubert Dreyfus had argued in his book "What Computers Can't Do"[25] that AI was fundamentally flawed because it was based on a flawed understanding of human intelligence. Dreyfus's book became a bestseller and helped popularize the view that AI was over-hyped and unlikely to succeed in the near future.

And so, the curtain fell on the first act of the artificial intelligence saga. Responding to the mood of the times, the flow of research funds from the US government, primarily through the Defense Advanced Research Projects Agency (DARPA), dried up. The reduction in funding fueled the perception that AI was over-hyped, creating a negative feedback loop.

The Rise of Expert Systems

Artificial intelligence is not a monolithic concept, though, and can mean other types of structures. Neural networks, particularly deep neural networks, are the most successful version yet. The 1970s through the 1990s were dominated by so-called expert systems as interest in neural networks waned.

These early AI models were designed to emulate the decision-making abilities of a human expert in a specific domain, such as medical diagnosis or financial planning. Expert systems were essentially decision trees, a series of if...then statements. They sought to capture and encode an expert's decision-making process so it could live as a program on a computer. Expert systems enjoyed considerable popularity until the dawn of the new millennium.

One example of an expert system is INTERNIST-I, a system to aid physicians in diagnosing medical problems. A team at the University of Pittsburgh developed a set of rules and heuristics from medical textbooks and consulted with medical experts. It started with formalizing the decision-making process of exactly one person, John D. Myers, M.D.[26] However, it never captured the confidence of the physicians it was supposed to help and largely remained a research tool. It was difficult to keep up with new knowledge, and coding it in. It also could not take a broad view of problems; the way it was structured could not consider a patient's history.

The finance industry today still uses expert systems. FICO, the credit analysis and fraud detection giant, uses an ingenious algorithm called RETE III, a rule-matching algorithm originally developed by computer scientist Charles Forgy.[27] Systems built on RETE III can efficiently sift through countless financial transactions, identifying patterns and trends that would be difficult for humans to detect, let alone at the algorithm's scale and speed.

AI Revival: A Fitful Resurgence

Artificial intelligence, as most commonly used today, is in the form of deep neural networks. *Neural* means a simulation of the function of the neural cells in our brains. From breathing to planning our next meal, training our bodies to lift heavy objects, and recognizing sounds, everything we do happens because of the way neurons in our brains are structured and communicate with each other. The emergence of these experiences from relatively simple electrical connections between neurons is still a matter of intense research. Still, researchers have accepted that it works and hope to achieve something similar by simulating that structure and function.

They build networks of artificial neurons, trying to create a simplified model of what they see in the brain. Admittedly, we're like a child watching construction workers build a house, mimicking their actions by attaching two pieces of wood to a plank and proudly declaring it a home.

The term "deep" means there are multiple layers within the network. Think of a shallow pool with just a few layers of water compared to a deep ocean teeming with life at various depths. In AI, a shallow network may have few layers (maybe even just one or two) while a deep network may boast many more. The perceptron we mentioned earlier is a shallow neural network containing just one layer.

The concept of artificial intelligence, embodied by neural networks as proposed in Minsky and Papert's book *Perceptrons*, experienced a revival with the 1986 publication of a groundbreaking paper in the journal *Nature*. Titled "Learning representations by back-propagating errors,"[28] this seminal work was authored by three visionary research psychologists: David E. Rumelhart, Geoffrey E. Hinton, and Ronald J. Williams.

They were visionary in the sense that they worked in the field of neural networks when it was dismissed and not seen as a fit subject for proper scientists. In fact,

Hinton, as a graduate student in cognitive psychology, was advised to explore neural networks only in his spare time. He persisted, and after defending his doctoral thesis on vision, he continued to work on neural networks.

Since those early days, Hinton has become a towering figure in the world of artificial intelligence. He is associated with many of the major breakthroughs in the field, earning him the moniker "Godfather of deep learning." Hinton wears that mantle well, even as he has sounded warnings about the pace and direction of recent developments in AI.[29,30] He comes from a family with an illustrious lineage in science. His great-great-grandfather was George Boole, who invented Boolean logic, the mathematics underpinning computer circuit construction. He is also related to the 19th century explorer Sir George Everest.

Hinton credited the invention of backpropagation to David Rumelhart (also a psychologist) after the publication of the paper in *Nature*. Backpropagation is a way of teaching neural networks. It was simple enough to apply to large and complex neural networks, and it remains the foundation of neural network training today. To conceptualize how it works, consider an analogy of a group of musicians tuning their instruments in an orchestra. When a musician plays a note that is in tune, the conductor nods in approval. If the note is out of tune, the conductor provides guidance on whether the note is too high or too low, helping the musician make the necessary adjustments. This process continues until all the instruments are in harmony, much like how backpropagation refines the neural network's performance.

Often the case in science, Hinton, Rumelhart, and Williams did not develop with these ideas in isolation. The ideas had been formulated by other researchers,[24] but their paper in *Nature* inspired and catalyzed the field. Backpropagation reignited widespread interest in neural networks because it was easy to understand, simple to implement, and computationally efficient. Despite these advantages, training large neural networks remained a prohibitive task in terms of time and money. This hints as to why neural networks still did not come to the forefront of AI. Once again, despite enormous advances in CPUs, the techniques proposed by the three mathematical psychologists were outpacing the day's technology.

Following this success, the nineties were marked by more remarkable developments in neural network research. New computational techniques were developed. One of the pioneers in this period was one of Hinton's students, Yann LeCun. Working against the tide in computer science research, where conventional wisdom still regarded artificial intelligence as a topic leading nowhere, he was responsible for some major breakthroughs. He built some of the first AI models called Convolutional Neural Networks, or CNNs, and taught them to recognize handwritten digits, a technological feat soon adopted by banks to read checks.[31] Eventually, LeCun and his colleagues Corinna Cortes from Google and Christopher Burges from Microsoft created one of the most storied datasets in modern AI, the MNIST database of handwritten digits.[32] It is almost a rite of passage for AI researchers learning the basics of artificial intelligence to creating a neural network model, replicating LeCun's work, that can recognize handwritten digits with high accuracy. The sense of mastery from this task propels them on to greater discoveries.

No account of the history of artificial intelligence would be complete without acknowledging the third member of the triumvirate that has indelibly shaped the field. Along with Hinton and LeCun, Yoshua Bengio, a Canadian originally from France, born of Moroccan Jewish heritage, jump-started AI and continues to guide the field even today. In 2000, his groundbreaking paper[33] demonstrated how neural networks could understand language, even with all the multifarious ways of expressing meanings. Bengio has pushed the boundaries of what was possible in AI and is responsible for many of the breakthroughs that have become tools of the trade for AI researchers today. In 2018, in recognition of their groundbreaking contributions to artificial intelligence, Yoshua Bengio, Geoffrey Hinton, and Yann LeCun were jointly honored with what is informally known as the "Nobel Prize of Computing," the prestigious Turing Award.

Figure 1.5 From left to right: Yann LeCun, Geoffrey Hinton, Yoshua Bengio. As imagined by Midjourney, prompted by the author.

The Birth of Modern AI

Then came a time when processors became fast enough, and memory became cheap enough that complex, deep learning neural networks could be trained. This was in the first and second decades of the new century. In 2012, the field experienced a sea change when Big Tech started making huge bets on AI. Once again, the Godfather of deep learning, George Hinton, was a central figure in this development.

Greater developments in neural network research marked the 1990s and the 2000s. New computational techniques were developed. The term big data became voguish, followed soon by machine learning. Techniques were developed that could glean insights from reams of data using advanced statistical techniques. Some of these techniques were applied to the problem of computation-hungry neural networks.

George Hinton and two of his students at the University of Toronto, Alex Krizhevsky and Ilya Sutskever, built a model called AlexNet that could identify the content in images. It was a powerful demonstration of what computing power could finally achieve. It was also an anticipated progression. The trio were responding to an annual challenge called the ImageNet Large Scale Visual Recognition Challenge, organized by Professor Fei-Fei Li and her team at Stanford University.

Fei-Fei Li is another person who stands out as a mover in the field. Growing up in China during the Cultural Revolution, her father imbued in her a love for photography, and she has described her passion as a way to connect with the natural world and understand its complexities. At Princeton, she understood that the field was ripe for neural network models, and large-scale datasets were needed to help researchers move forward. Therefore, she developed ImageNet, a database of digital images which now holds 14 million images across 1,000 categories. The profound influence of ImageNet is evident as legions of AI researchers have honed their skills, drawing invaluable insights from this repository. Like LeCun's MNIST database, ImageNet has become one of the most consequential objects in the field of AI.[34,35]

The year 2012, the year AlexNet took the world by storm, was the first year neural network models were used to classify the images in ImageNet. Hinton and his students founded a company called DNNresearch. The large tech companies, sharklike instincts immediately recognizing the practical power of their algorithm, descended on Lake Tahoe in December 2012. There, in a casino hotel room, they engaged in a tense auction for the fledgling three-person company. This was a high-stakes secret auction, and the company representatives soon found themselves bidding outside their authorized limits. Phone calls and emails to headquarters to sign off on ever-increasing bids resulted throughout the day, and finally, one company walked off with the prize. The amount of the winning bid was never officially disclosed. Still, there were reports in the media that Hinton shut down the auction at $44 million[36] despite every indication that the amount could go higher. Hinton later stated, tongue in cheek, "I signed contracts saying I would never reveal who we talked to. I signed one with

Microsoft and one with Baidu and one with Google." Hinton and his students joined Google soon after.

AlexNet was one of many successful image recognition models. A couple of years earlier, DanNet, developed by Dan Ciseran, had won other competitions as well. It was a more specialized algorithm focusing on medical images. What AlexNet, DanNet, and other algorithms made clear is that the age of deep learning had arrived, and an arms race had begun among the top tech companies.

AI Today

This inter-company arms race would get much more heated 10 years later when the OpenAI consortium unleashed ChatGPT on the world. This technology, which seems to be paving the way for the incredible spoken computer interface in the TV series *Star Trek*, has delighted the public with its oracle-like abilities, similar to the delight that Google search occasioned in the naughts. But the amount of research it enables is balanced by the amount of misinformation it blithely generates.

AI's ability to understand natural language had already up-ended the translation industry. In yet another science-fiction foreshadowing, the babel fish from *The Hitchhiker's Guide to the Galaxy* no longer seems like a ridiculous, incredible device. In the book, the babel fish, when inserted into your ear, would immediately translate any language to English or your preferred language. For tourists around the world, Google's Translate app performs that very function in a more rudimentary way.

For several years, research in speech and language understanding continued, but at a different rate than computer vision. GPT-3.5 was truly an eruption in this landscape, an achievement that caught even the AI cognoscenti by surprise. It is part of a new trend in AI research, the creation of foundational models that can be applied to many situations. Other foundational models are ones that generate images, such as DALL-E 2 by OpenAI, and Midjourney. Foundational models are being built for scientists, such as Segment Anything by Meta[37] that enable AI programmers to quickly find objects in any image.

What this means is that the ability to harness AI is being opened up to more people. These models are generally available to anyone right now. But no doubt they will be part of the AI-powered economy of the 21st century.

Driver of the 21st Century Economy

AI, specifically connectionist AI, has evolved from a nascent concept into a juggernaut within just 20 years. Trillion-dollar companies like Microsoft and Google have recognized the financial potential of AI and have pivoted their businesses to integrate AI into every aspect of their operations.

Microsoft's Azure, for instance, has grown its cloud computing platform into an AI-driven powerhouse, enabling businesses to harness the power of AI and machine learning for their applications. Google has also invested heavily in AI research and development, with projects like TensorFlow and Google Brain propelling the company to AI dominance.

Meta, formerly known as Facebook, also made a bold move by shifting its focus to an AI-powered metaverse. This ambitious endeavor has seen the company take on losses of more than $10 billion in order to develop its new business model. The metaverse, a virtual reality space where users can interact with a computer-generated environment and other users, was expected to be a game-changer for businesses. But disappointing results have led them to change their business: a de-emphasis on the metaverse and doubling down on AI.

The widespread adoption of AI across industries has also profoundly impacted the job market. Job descriptions now commonly mandate a requisite understanding of AI. Employers appreciate the worth of employees who possess the ability to navigate new technologies, even if such employers may not comprehend the technology themselves. Entirely new job roles have arisen, such as AI ethicists, who are responsible for ensuring that AI systems are designed in a manner that respects human values and ethical principles. Universities and online platforms have reacted to the surge in demand for AI-related skills by offering specialized courses to equip professionals with the necessary expertise.

It is no secret that big tech companies have been on a talent acquisition spree, snapping up the brightest minds in AI research and development. Google, for instance, has made significant strides in consolidating its position in the AI space, reportedly absorbing 30 AI startups at a cost of around $4 billion.[38] This aggressive expansion led to concerns that the available AI talent pool is being monopolized by the tech giants, potentially stifling innovation and leaving little room for newcomers to thrive.

There has been a surprising surge in global funding for AI startups. The exponential increase from $4.5 billion in 2014 to a staggering $38 billion[39] in just the first half of 2021 suggests continued international investor interest and capacity. Yet it also left many wondering if this is an investment bubble driven by the scarcity of AI knowledge as an essential commodity and fear of missing out on the next big breakthrough in AI or if it is a genuine testament to the financial potential of AI. However, the Great Layoff of 2023 released a considerable depth of AI talent into the market, which should fuel another surge of AI startups and funding.

The global nature of AI funding highlights the fact that AI innovation is not limited to Silicon Valley or other established tech hubs. Emerging markets in Asia, Europe, and beyond are increasingly becoming hotbeds for AI research and development. In some countries, such as China, AI is rapidly becoming part of social government, while in Europe, politicians are looking to curb and limit intrusions AI might make on privacy. The impacts of these different approaches will be explored in Chapter 6.

This amount of migration to AI, of both people and money, must be happening because investors hope to make large profits. Sizing market potential is a necessary step for businesses, but this task is far from a straightforward scientific endeavor. In fact, it is more of an art, a combination of intuition, experience, and data analysis. Companies such as McKinsey have made billions of dollars by mastering that art or at least convincing the world that they have.

In 2018, McKinsey[40] released a report on the value of AI, estimating that it could range between $3.5 trillion and $9.8 trillion by 2030. This staggering range is a testament to the complexity and uncertainty that characterizes market sizing. PricewaterhouseCoopers,[41] one of McKinsey's competitors, put the figure at an even more astounding $15.7 trillion.

These impressive numbers may turn out to be wildly inaccurate, but they strive to capture the potential of AI to revolutionize virtually every aspect of our lives. The technology has already shown remarkable promise in areas such as natural language processing, image recognition, and predictive analytics, to name just a few. And our successes in these areas have opened a window into what's possible in the future.

For instance, as we grapple with the repercussions of climate change, we will find AI invaluable in finding solutions for improved forecasting and disaster management. In the area of health, AI-powered medicine could significantly improve the accuracy and speed of diagnoses and prescribe personalized and more effective treatments for patients. In the finance field, AI could enable more sophisticated fraud detection and nuanced risk analysis. Drones, self-driving trucks, and cars could transform the way we move goods and people, improving safety and environmental health.

But the darker side of AI is waiting. If we flip the AI coin, we will see AI powering devices and machines that are not beneficial for society. We cannot wish away concerns by George Hinton that the technology he was such an instrumental part of may end up being detrimental to humanity. AI can be used to spread misinformation, which can cost lives. It can amplify the worst of human society, like racism and sexism, wealth inequalities, and health inequity, because it learns from human society. What about the loss of human judgment and critical thinking? If AI is writing the news and creating music and art, what does it mean to be human? We will explore some of these issues in later chapters.

Scientists and politicians recognize these risks and are creating guidelines, standards, and laws that will aim to mitigate the negative consequences of this new technology. The more people are aware of AI's abilities and limitations, what it can and cannot do, the better our collective use of AI will be. AI literacy will be as important as traditional literacy and numeracy in making informed choices about how we will live with AI.

AI is not just another consumer product, a gadget on the shelf. Nor is it a mere extension of cloud services or another way to compute like Azure or Amazon Web Services (AWS). At its core, AI is a collection of algorithms and computational models designed to mimic the cognitive abilities of the human brain. By processing vast amounts of data, these algorithms can identify patterns and make correlations, thereby allowing machines to "learn" and adapt to new information. This ability to process and analyze data at lightning speed is what sets AI apart from traditional computing. The true essence of AI lies in its ability to enable humans to make efficient use of resources, a capacity that we commonly refer to as intelligence. In this sense, it is an intangible force that empowers us to coexist with the world and manage its resources in once unimaginable ways.

Final Thoughts

Historians caution that the tides of human progress are caused by factors much larger than individuals. There are broad societal and cultural forces that guide our energies and shape the direction of our progress and occasional regress. However, in narrating the past, certain personalities stand out as bright representatives of their times, shining light on their eras. We should also remember that countless other souls fueled their breakthroughs in thought and understanding.

The ebbs and flows of AI, its near emergence and temporary declines, yet a consistent upward trend that is clear in hindsight, mirrors that of other technologies. The parallel leads us to expect challenges we can barely imagine now. Let us hope that humanity chooses to use this technology for the good of our world.

References

1. Kasparov, G. K. & Greengard, M. *Deep Thinking: Where Machine Intelligence Ends and Human Creativity Begins.* (PublicAffairs, an imprint of Perseus Books, LLC, 2017).
2. Hsu, F. *Behind Deep Blue: Building the Computer that Defeated the World Chess Champion.* (Princeton University Press, 2004).
3. Google CEO: AI impact to be more profound than discovery of fire, electricity — CBS News. at <https://www.cbsnews.com/video/google-ceo-ai-impact-to-be-more-profound-than-discovery-of-fire-electricity/>
4. Gates, B. The Age of AI has begun. *gatesnotes.com* at <https://www.gatesnotes.com/The-Age-of-AI-Has-Begun>

5. Clifford, C. Elon Musk: "Mark my words — A.I. is far more dangerous than nukes." *CNBC* (2018). at <https://www.cnbc.com/2018/03/13/elon-musk-at-sxsw-a-i-is-more-dangerous-than-nuclear-weapons.html>
6. Swade, D. *The Cogwheel Brain: Charles Babbage and the Quest to Build the First Computer*. (Little, Brown, 2000).
7. Hyman, A. *Charles Babbage: Pioneer of the Computer*. (Princeton University Press, 1985).
8. Campbell-Kelly, M. Charles Babbage's Table of Logarithms (1827). *Annals of the History of Computing* **10**, 159–169 (1988).
9. Babbage, C. *Passages from the Life of a Philosopher*. (2018). at <https://www.gutenberg.org/ebooks/57532>
10. Lethbridge, L. *Ada Lovelace: Computer Wizard of Victorian England*. (Short Books, Limited, 2004).
11. Essinger, J. *Ada's Algorithm. How Lord Byron's Daughter Ada Lovelace Launched the Digital Age*. (Melville House Books, 2013).
12. Macrae, N. *John von Neumann: The Scientific Genius Who Pioneered the Modern Computer, Game Theory, Nuclear Deterrence, and Much More*. (Plunkett Lake Press, 2019).
13. Bhattacharya, A. *The Man from the Future: The Visionary Ideas of John von Neumann*. (W. W. Norton, 2022).
14. von Neumann, J. & Kurzweil, R. *The Computer and the Brain*. (Yale University Press, 2012).
15. Turing, A. M. *Computing Machinery and Intelligence*. (Springer, 2009).
16. Turing, A. & Copeland, B. J. On Computable Numbers, With An Application To The Entscheidungsproblem. in *The essential Turing: seminal writings in computing, logic, philosophy, artificial intelligence, and artificial life, plus the secrets of Enigma* (Clarendon Press; Oxford University Press, 2004).
17. McCarthy, J., Minsky, M. L., Rochester, N. & Shannon, C. E. A Proposal for the Dartmouth Summer Research Project on Artificial Intelligence, August 31, 1955. *AI Magazine* **27**, 12 (2006).

18. The Dartmouth Workshop—as planned and as it happened. at <http://www-formal.stanford.edu/jmc/slides/dartmouth/dartmouth/node1.html>
19. Dartmouth AI Archives. at <https://raysolomonoff.com/dartmouth/>
20. Rosenblatt, F. The perceptron: A probabilistic model for information storage and organization in the brain. *Psychological Review* **65,** 386–408 (1958).
21. Minsky, M. & Papert, S. A. *Perceptrons: An Introduction to Computational Geometry*. (The MIT Press, 2017). doi: 10.7551/mitpress/11301.001.0001
22. Darrach, B. Meet Shaky, the first electronic person. *Life* **69,** 58B – 68 (1970).
23. Simon, H. A. *The New Science of Management Decision*. (Prentice-Hall, 1977).
24. Russell, S. J., Norvig, P. & Davis, E. *Artificial Intelligence: A Modern Approach*. (Pearson India Education Services Pvt. Limited, 2015).
25. Dreyfus, H. L. *What Computers Can't Do: A Critique of Artificial Reason*. (Harper & Row, 1972).
26. Miller, R. A., Pople, H. E. & Myers, J. D. in *Computer-Assisted Medical Decision Making* (eds. Reggia, J. A. & Tuhrim, S.) 139–158 (Springer, 1985). doi:10.1007/978-1-4612-5108-8_8
27. Forgy, C. L. Rete: A fast algorithm for the man pattern/many object pattern match problem. *Artificial Intelligence* **19,** 17–37 (1982).
28. Rumelhart, D. E., Hinton, G. E. & Williams, R. J. Learning representations by back-propagating errors. *Nature* **323,** 533–536 (1986).
29. Metz, C. "The Godfather of A.I." Leaves Google and warns of danger ahead. *The New York Times* (2023). at <https://www.nytimes.com/2023/05/01/technology/ai-google-chatbot-engineer-quits-hinton.html>

30. Tavernise, S. The Godfather of A.I. has some regrets. At <https://www.nytimes.com/2023/05/30/podcasts/the-daily/chatgpt-hinton-ai.html>
31. LeCun, Y., Bottou, L. & Bengio, Y. Reading checks with multilayer graph transformer networks. In *1997 IEEE International Conference on Acoustics, Speech, and Signal Processing* **1**, 151–154 (IEEE, 1997).
32. LeCun, Y., Cortes, C. & Burges, C. *MNIST handwritten digit database*. (Florham Park, NJ, 2010).
33. Bengio, Y., Ducharme, R. & Vincent, P. A neural probabilistic language model. *Advances in Neural Information Processing Systems* **13** (2000).
34. Li, F. F. *The Worlds I See: Curiosity, Exploration, and Discovery at the Dawn of AI*. (Flatiron Books, 2023).
35. Kaplan, J. *The Genius of Women: From Overlooked to Changing the World*. (Penguin Publishing Group, 2021).
36. The Secret Auction That Set Off the Race for AI Supremacy | WIRED. at <https://www.wired.com/story/secret-auction-race-ai-supremacy-google-microsoft-baidu/>
37. Kirillov, A., Mintun, E., Ravi, N., Mao, H., Rolland, C., Gustafson, L., Xiao, T., Whitehead, S., Berg, A. C., Lo, W.-Y., Dollár, P. & Girshick, R. Segment Anything. Preprint at https://doi.org/10.48550/arXiv.2304.02643 (2023)
38. Hurst, A. Google revealed to have acquired the most AI startups since 2009. *Information Age* (2020). At <https://www.information-age.com/google-revealed-acquired-most-ai-startups-since-2009-15415/>
39. CB Insights. Total funding of AI startups worldwide 2014-2021. *Statista* at <https://www.statista.com/statistics/621468/worldwide-artificial-intelligence-startup-company-funding-by-year/>

40. Sizing the potential value of AI and advanced analytics | McKinsey. At<https://www.mckinsey.com/featured-insights/artificial-intelligence/notes-from-the-ai-frontier-applications-and-value-of-deep-learning>

41. PricewaterhouseCoopers. PwC's artificial intelligence services. *PwC* (2023). at <https://www.pwc.com/us/en/services/consulting/cloud-digital/data-analytics/artificial-intelligence.html>

2

AI Explained: A Non-Technical Guide

"Every man can, if he so desires, become the sculptor of his own brain."

– Santiago Ramón y Cajal

"Every act of perception, is to some degree an act of creation, and every act of memory is to some degree an act of imagination."

– Oliver Sacks

In this chapter, we will explore the lexicon that has developed around AI—the shared language that has grown out of its history, and the vagaries of its practitioners. The cant employed by specialists may seem mysterious; however, by drawing parallels to universally shared experiences, we can make the mysteries of the field more accessible.

Definition of AI

Let's start with a type of artificial intelligence called *expert systems*. We will not be addressing this type of AI in much detail because it is not at the center of the current resurgence. Expert systems in AI had its heyday during the AI winter of the 1970s and were the major focus of development in the subsequent decades. The researchers and engineers building these systems believed they could encapsulate the decision-making processes of an expert in the field. It seemed reasonable to expect that if an expert could explain how they reached a decision step by step, that process could be encoded in a computer program. Assuming experts made their decisions logically, then logical rules, commonly in the basic format of if. . .then, could reasonably be expected to replicate an expert. A bank loan officer, for example, could say they looked at an applicant's credit score, the totality of their assets and debts, and other data. These can be fed into a computer, crunched by an algorithm, and a credit score indicating risk would be the output.

Whereas machine learning neural networks mimic the neural function of the human brain, expert systems are designed to mimic the higher-level decision-making abilities of human experts. They are usually positioned as decision aids rather than as total replacements for human jobs. The loan officer is often free to say, "I looked at the expert systems' recommendation, but I also place importance on my gut feel of the applicant when they are sitting in front of me."

Expert systems played a role in medical and clinical decision-making as well. We saw two examples in Chapter 1. They follow similar principles like the clinical decision support system called Dxplain, which is used at Massachusetts General Hospital. Physicians, particularly medical students, can input data about their patients, such as their symptoms and lab results. This data makes their way through decision trees and sets of rules, generating a ranked list of possible diagnoses. You might have encountered simpler versions of expert systems on popular medical websites like WebMD. By entering your age, sex, medications, and symptoms, you can receive a list of potential, often alarming diagnoses. While this might grab your attention, it is important to approach expert systems with caution. They can be accurate and effective but require extensive domain-specific knowledge to develop.

Furthermore, uncertainty or incomplete information can hinder their effectiveness.

Expert systems are also extensively utilized in credit scoring systems, such as FICO, familiar to anyone who has tried to buy a house in the United States. These systems incorporate a number of tools, including machine learning. But they can also use decision trees, essentially flowcharts of if. . .then statements, acting on a variety of input variables such as the loan applicant's income, current debt, history of repayments, and so on. Through these mechanisms, credit scoring systems provide a final assessment of the risk associated with a new loan application.

Machine Learning

Expert systems were overshadowed in the first decade of this century by a group of techniques called *machine learning*. The term originated at IBM in the 1950s, and credit for its coining belongs to a computer scientist named Arthur Samuel.[1]

Samuel built a computer program that could play checkers with a human opponent. What fascinated scientists was the program's ability to learn from its mistakes, a big jump from set-in-stone instructions that were the norm. The program appeared on national TV in 1956, foreshadowing IBM's Deep Blue's battle with Garry Kasparov 40 years later and IBM's Watson TV appearance on the trivia game show *Jeopardy*. Like its chess-playing descendant, it beat its human opponent and contributed to the optimism about artificial intelligence at the time. The fact that it lost several subsequent games did little to curb the enthusiasm.

Machine learning is considered a subset of artificial intelligence. One of the most popular approaches to machine learning is *neural networks*, which are a type of model inspired by the structure and function of the human brain. As the name implies, neural networks consist of interconnected nodes or neurons. The connections between them may be strong or weak, allowing information to pass between them easily or with difficulty.

While all neural networks are a form of machine learning, not all machine learning techniques are neural networks. If one thinks of machine learning as a toolbox for artificial intelligence scientists, then neural networks are but one tool in there. Two other tools that may be familiar to readers, out of the dozens in there, are decision trees and support vector machines (SVMs). They are statistical models that do not have the same structure as neural networks and process data very differently.

One of these tools is called a *decision tree*. Think of it as a flowchart-like structure that helps the computer make decisions by following a series of logical paths. It breaks down a problem into smaller, simpler questions and makes choices based on the answers. It is like asking a series of yes-or-no questions until you reach a conclusion. SVMs use a different algorithm. This algorithm tries to find the best line, which could be a squiggly line, to separate as much of one group from another. For example, in a satellite image, it can be used to separate a tilled field from a forest.

There are several other tools like this in the toolbox, with names like *linear regression* and *K-nearest neighbors* (KNN), each with their own way of data input, processing, and output. Linear regression is used for prediction.

It looks for correlations between variables and extrapolates that relationship to make predictions. For example, age and dementia are correlated, so a regression model would give you the odds of dementia at different ages. KNN is a way to classify new data into predetermined categories. It looks at the attributes of a new object and determines how many of them are most similar to any existing categories.

With the advent of big data and more powerful computers in the late 2000s, machine learning became a must-have for tech companies. Techniques that were once theoretical could now be implemented, enabling computers to learn from data and improve their performance on a task without being explicitly programmed.

How Machines Learn

Animals, including Homo Sapiens, learn by associating cause and effect. If a toddler sees a bunny and says "cat!," they will be corrected by their wiser preschooler sibling. Then, the toddler is likely to refer to subsequent sightings of the animal not as a cat but something resembling the sound "bunny." In school, we may memorize things by rote or we may learn through experience. In psychology, learning is the process by which we remember. These memories may be recalled automatically, such as when we ride a bike and our muscles know what to do, or consciously, as when recalling that 6 times 10 is 60.[2,3]

Neural networks also rely on associative learning to recognize patterns and make predictions. Like animal brains, neural networks also learn through experience. This process involves learning the associations between certain inputs and outputs. For example, a neural network trained to identify images of cats will learn to associate specific features of an image, such as the shape of the ears or the color of the fur, with the label "cat."

The process of associative learning in neural networks is similar to how humans learn. Both involve the formation of connections between different pieces of information. In neural networks, these connections are represented by the weights between neurons. As the network is trained, these weights are adjusted to strengthen the connections between inputs and outputs associated with the correct label. In the brain, these connections are the strength of the electrical connection or how much effect the firing of one neuron has on another. In the process of learning, these strengths may increase or decrease.

Emotion also plays a role in the learning process for both humans and neural networks.[4] It has been well established, through human experience and psychological research, that emotion of some sort helps sustain the learning in our brains. We may remember an event that was particularly emotionally charged more vividly than something that was mundane. We also forget. Sherlock Holmes famously compared his mind to an attic with only so much room, so when some information was no longer needed, it would be pushed out. The brain does not quite work this way, but we do forget unintentionally. Or, we may have the memory still encoded in our brains but cannot retrieve it. This leads to phenomena such as tip-of-the-tongue phenomenon,[5] or blanking on someone's name, which are universally experienced.

When we experience emotions such as fear or pleasure, the brain releases neurotransmitters that strengthen the connections between neurons. Likewise, in neural networks, the strength of connections between neurons increases when the network is rewarded or penalized based on accurate or inaccurate predictions.

Just like humans, neural networks are also prone to forgetting. A common problem in neural networks was known as catastrophic forgetting.[6] This happened when a neural network was trained on a new set of data, occasionally causing it to forget the associations it had learned from a previous dataset.

How have artificial neural networks emulated all these aspects of human neural networks? We shall explore that by considering three major types of learning.

Supervised Learning

Supervised learning is a type of machine learning in which the model is trained on labeled data. This means that the input data is paired with corresponding output data that serves as a "correct" answer. A direct analogy is a teacher giving a student a set of study questions and answers before an exam. The teacher provides the correct answers, and the student learns by comparing their answers to the correct ones. In the same way, a supervised learning algorithm is trained on a set of labeled data, where the correct answers are provided and the algorithm adjusts its parameters to improve performance.

In practice, during training, the weights between the neurons in a model are adjusted to minimize the difference between the model's predicted outputs and the true outputs in the labeled data. These weights are analogous to the strength of the connection between neurons in the brain.

Think of the adjustment of these weights as turning a knob to fine-tune a model's output. Eventually, you settle on a combination of weights so that the output of the model is reasonably close to the input. That is, if you show a picture of a cat, the output may be text that says, "cat."

The weight adjustment procedure is repeated many times in a systematic way. But it is computationally taxing. With today's technology, hardware consisting of fast GPUs and programming languages designed to translate a programmer's intentions to computerese, the time it takes to arrive at a suitable configuration of weights could take hours, days, or even weeks.

Once the weights have been set to the machine learning scientist's satisfaction, the model is said to have been trained. Additional training can happen if there is new labeled data. In principle, the more labeled data one throws at a supervised learning algorithm, the more accurate it will be.

Supervised learning is used in a wide range of applications, including image and speech recognition, natural language processing, and recommendation systems. If there is sufficient training data, then supervised learning is a good approach.

Unsupervised Learning

Unsupervised learning differs from supervised learning in that the algorithm does not have answers when it is being trained. In the earlier example of the teacher giving students both the questions and answers, the unsupervised learning students would only get the questions. It is arguably a much harder way to learn.

As another example, consider a group of people at a party who have not met before. Over time, they start to notice similarities and differences between them. There will be people who react positively to remarks on movies, and other people who gather close to the bar. They will form their own groups based on shared interests or personality traits. In the same way, an unsupervised learning algorithm is given a set of unlabeled data and tries to find patterns and structure within the data on its own.

This means that the input data does not have any corresponding output data that serves as a guide. An interesting aspect of unsupervised learning is that it may find commonalities and patterns among the data that humans would not have considered. In medical image scans, for example, these algorithms may be able to detect important diagnostic features that even experts may not have caught on to. Unsupervised learning has also been used to create AIs that can emulate an artist's style and produce new artwork. For example, an AI being trained via unsupervised learning will learn the features of Picasso's cubist style and will render a photograph of your spouse as a potential modern masterpiece in the artist's style. The AI has presumably learned what characterizes Picasso's art, such as the colors, the lines, the relative positions of facial features, etc.

Within the context of artificial neural networks, unsupervised learning can be more laborious than supervised learning. When the data is unlabeled, more of it is needed for the models to make some sort of useful representation. But labeling data is also laborious and expensive, especially in the amount needed to train a model, as it requires manual labeling by human experts. On the other hand, unlabeled data is much more abundant and can be collected easily. Researchers have found that they can greatly speed up the learning process if they give the model even a little direction. Enter semi-supervised learning, which uses the large amount of unlabeled data to improve the performance of the model, while still relying on the smaller amount of labeled data to provide guidance. An example of this approach is speech recognition. The unsupervised part creates features from audio and sends these features to a second model, a supervised learning model, that can classify the features into categories. For example, it can detect which language the audio is in, or it can decide whether the audio is angry or happy.

Reinforcement Learning

Reinforcement learning stands as the foremost variant of learning, one that permeates all our lives. A student learns by interacting with the environment and receives feedback from a teacher in the form of rewards or punishments—gold star stickers or less time for recess. Animal behaviorists are very familiar with this type of training. A dolphin can be trained to jump through a hoop or retrieve objects from the bottom of a deep pool by being rewarded when it performs the action successfully. And it is punished when it does not perform that action, which can be as mild as withholding a reward. Over time, the dolphin learns to perform the tricks more reliably. In the same way, a reinforcement learning algorithm interacts with an environment and receives rewards or punishments based on its actions. The algorithm learns to make decisions that maximize its rewards over time.

Psychologists pioneered the study of reinforcement learning in a quest to understand behavior. Burrhus Frederic (BF) Skinner, an American researcher and one of the originators of the field of behaviorism, believed the environment shaped behavior completely.[7] His predecessor John B. Watson (Figure 2.1) said, "Give me a dozen healthy infants, well-formed, and my own specified world to bring them up in, and I'll guarantee to take any one at random and train him to become any type of specialist I might select — doctor, lawyer, artist, merchant-chief, and, yes, even beggar-man and thief, regardless of his talents, penchants, tendencies, abilities, vocations, and race of his ancestors."[8] Skinner built his theories on a similar belief and developed a technique called operant conditioning that examined how reinforcement could shape any living thing's actions. Research in this area varies the types of rewards, the time between rewards, etc., to determine the best methods of training novel behaviors. Current research paths take into account not just outwardly observed behaviors but also predilections based on genetics and cognitive processes, which may not be readily apparent.

The goal of reinforcement learning is to learn a policy or strategy that maximizes the expected cumulative reward over time. This field has drawn insights from psychological research on optimal learning strategies and has been instrumental in creating autonomous agents capable of making decisions in complex environments. AlphaGo, an AI that plays the game of GO, and developed by the company DeepMind, is the result of reinforcement learning.[9] Go is a complex board game with an enormous number of possible moves, making it

Figure 2.1 John Watson conditioning Little Albert to be afraid of furry creatures and similar objects, even Santa Claus!

Credit: Screenshot from movie on Wikimedia common. Original video posted on YouTube by Jaap van der Steen, Baby Albert Experiments - YouTube.

challenging for traditional brute-force search methods to find optimal solutions. AlphaGo, however, employed a combination of deep neural networks and reinforcement learning techniques to master the game. In the training process, AlphaGo initially learned from a large dataset of expert human games, acquiring a foundation of strategic knowledge. Subsequently, it used reinforcement learning to further improve its gameplay. AlphaGo played numerous games against itself, with each move evaluated based on the potential future rewards it could lead to.

Through this iterative process, the AI system honed its skills, came up with strategies not seen before in the game, and eventually beat world champion Lee Sodel.[10]

Neural Networks

The human brain has anywhere between 80 and 90 billion cells called neurons.[11] They are connected to each other in electrical pathways, forming neural networks. These networks are responsible for all of our cognitive abilities. They form our thoughts and feelings; they help us breathe, our hearts beat, and every part of our body to function in concert. They are responsible in some as yet unknown way for our experience of consciousness. They achieve this by sending electrical signals to each other via established and (as we've recently discovered) ever-evolving networks. If you want to lift a finger, a particular sequence of neuronal electrical firings will activate the relevant muscles to perform that action. Take the miracle of sight, which is somehow achieved by the lobes at the back of our brains. When light enters the retina and goes through the optic nerve to the back of the brain, it sets off a cascade of signals and we experience the sight of a cat, the color green, or the sense of depth and distance between objects.[12]

To put it into perspective, imagine a massive city with millions of people. The neurons in our brain are like the individual citizens, and the neural networks they form are like the roads, highways, and transit systems that connect them. Just as the city would grind to a halt without efficient transportation, our brains wouldn't be able to function without these networks.

These networks can also break down, resulting in a range of neurological disorders. Malfunctioning in these networks can cause epilepsy, Alzheimer's, Parkinson's, and a whole host of other neural diseases. Other functions seem like drawbacks, but maybe only because we have not found good reasons. Why should we ever forget anything, and why do we need sleep?

Trying to understand these mechanisms is how psychologists like Frank Rosenblatt stumbled on to the creation of artificial neural networks. Let us explore how artificial neural networks work.

Structure of a Neuron

A neuron can be thought of as a light switch in a room. When the switch is flipped on, electricity flows through the wires and a light turns on. Similarly, when a neuron receives a strong enough signal from other neurons, it is switched on. For a neuron to be switched on means it fires a signal and sends an electrical current to other neurons. The strength of the signal can be adjusted by turning a dial on the switch, which is like adjusting the synaptic weights of the neuron.

The structure of the neuron was laid bare by the collective efforts of several scientists, but one who stands out was Santiago Ramón y Cajal. He was a Spanish scientist who studied the brain in the late 19th and early 20th centuries. Cajal used a type of stain called an aniline dye to study the structure of neurons, the same dye that is used for fibers and wood. By taking a very thin slice of brain tissue, putting it on a glass slide, and then adding a few drops of aniline dye, he could discern the typical branching structure of neurons. He discovered that neurons have a unique shape and structure that allow them to communicate with each other. Since he also happened to be quite a good artist, his drawings of neurons and the structures they formed helped his discoveries find a receptive audience among neuroscientists (Figure 2.2).[13,14]

Ramón y Cajal showed that a neuron is the basic building block of the nervous system. It consists of three main parts: the dendrites, the cell body, and the axon.

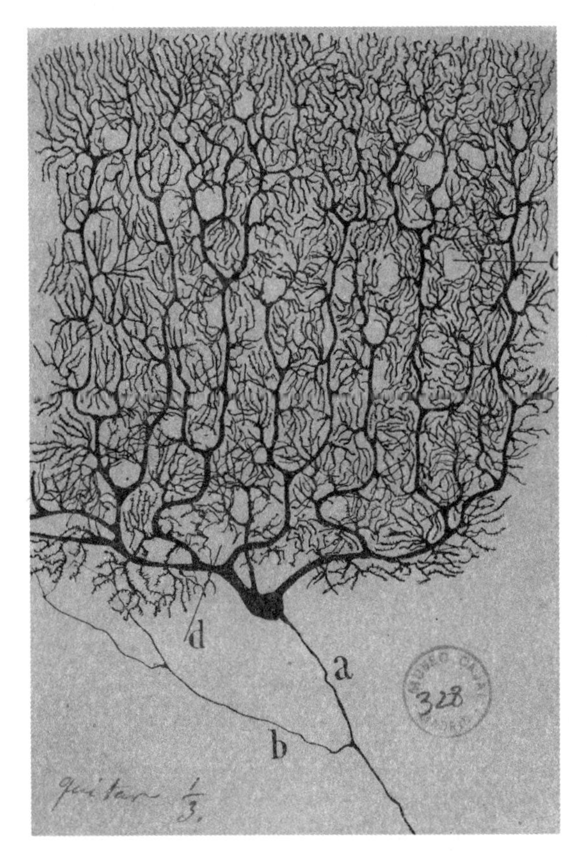

Figure 2.2 Ramon y Cajal's drawing of neurons in the cerebellum, 1899

Credit: Cajal Institute, Madrid

The dendrites receive signals from other neurons, which are transmitted to the cell body. If the signals are strong enough, the cell body sends an electrical signal down the axon, which can activate other neurons downstream. The strength of the signal can be adjusted by the neuron's synaptic weights, which determine how much weight to give to each incoming signal. By adjusting these weights, a neuron can learn to recognize certain patterns or features in the input data. Neuroscientists have since mapped the brain in great detail and uncovered several different types of structures that cells can exhibit.

The artificial neural networks we work with are much simpler. They consist of a single entity defined by whether it is on or off, one or zero, and further defined by the strength of its connections to other neurons around it. So, when we build artificial neural networks, we are emulating not their precise biological structure but just their function.

Layers in a Neural Network

Minsky and Papert suggested that arranging neurons in layers, and then having the layers communicate with each other, would lead to more powerful computations.[15] Indeed, this is how we arrive at deep learning—by having multiple layers of neurons or deep layers communicating with each other. This communication across layers is key to the rise of complex behavior.

To understand the importance of layers, let's think of a neural network as a team of detectives investigating a crime. The detectives in the first layer collect evidence at the crime scene, such as fingerprints and footprints. They pass this evidence on to the detectives in the second layer, who use it to identify potential suspects based on their characteristics. The detectives in the third layer then use this information to build a case against the most likely suspect. In the same way, each layer of a neural network learns to extract increasingly complex features from the input data until the final output is produced.

A neural network is composed of multiple layers of neurons that are interconnected in a specific way. The first layer, called the input layer, receives the raw input data, such as an image or a sound wave. The input data is then processed through a series of hidden layers, which learn to extract increasingly complex features from the data. Finally, the output layer produces the final output of the network, which can be a classification label, a prediction, or some other form of output. The number of hidden layers and the number of neurons in each layer can vary depending on the complexity of the problem and the amount of data available for training.

The importance of layers in artificial neural networks, although theorized for years, really took off thanks to Geoffrey Hinton and his collaborators, David Rumelhart and Ronald Williams. Their groundbreaking contribution was to develop an algorithm, which was capable of performing the heavy computations necessary for setting the weights between neurons. As discussed in Chapter 1, they developed a remarkable technique called backpropagation, which allowed neural networks to learn from input data and dynamically adjust their synaptic weights.[16] This breakthrough led to the practical development of multilayer neural networks, which are now widely used for tasks such as image and speech recognition. In the next section, you will see how backpropagation works.

Training a Neural Network

Training a neural network can be thought of as teaching a child to ride a bike. At first, the child may wobble and fall, but with practice, they learn to balance and steer the bike. In the same way, a neural network starts with random synaptic weights and makes many mistakes at first. But with each iteration of training, the network adjusts its weights to reduce its errors and improve its performance. Like a child learning to ride a bike, the network gets better with practice until it can make accurate predictions on new, unseen data. This practice, or learning, happens using the methods of machine learning described earlier. Supervised learning, unsupervised learning, and reinforcement learning can all be used to train neural networks.

Backpropagation: Learning from Mistakes

Training a neural network involves adjusting the synaptic weights of the neurons so that the network can learn to recognize patterns or make predictions based on the input data. This is typically done using a process called backpropagation.[17] To put it succinctly, backpropagation is like a feedback system that helps the AI learn and improve by gradually adjusting the connections between its neurons so it can recognize objects more accurately. The training process has to be repeated many times, using different subsets of the data for each iteration, until the network's performance is satisfactory. Once the network is trained, it can be used to make predictions on new, unseen data.

To give a quick illustration of how backpropagation works, let's take the example of an AI model that can look at pictures and describe what it sees.

If one were to give it an image of a sunflower, it should output the word "sunflower." To do this task, the AI is made up of an artificial neural network. This neural network consists of different layers of neurons that are connected to each other. In our case, the first layer of neurons takes the picture as input. However, the network does not understand pictures directly. Instead, it represents the picture as a bunch of numbers, which are like electronic versions of the image.

These numbers go through several layers in the neural network, with each layer performing some calculations. Eventually, the numbers reach the last layer, which gives us the output. The output here is still a list of numbers. To check if the AI's answer is correct, we compare its output (the list of numbers) with the numerical representation of the word "sunflower." If the two lists of numbers are different, it means the AI made a mistake.

This is where backpropagation comes in. When the AI's output is different from the expected answer, the backpropagation algorithm starts working. It goes backward through each layer of the neural network and all the connections between the neurons, and while traveling backwards, it makes small changes to the weights of these connections, which are like knobs that control the information flow. That is, it is propagating the error between the model's output and the desired output backwards through all the layers of the network. The goal is to improve the network's performance so that when we show the sunflower picture again, the difference between the AI's output and the correct answer becomes smaller.

This process keeps repeating until the difference between the neural network's output and the true answer becomes negligible, meaning it gets really close to the correct answer. This way, the neural network learns from its mistakes and becomes better at recognizing objects in images. If the backpropagation algorithm sounds laborious, it certainly is in practice. This is why powerful chips had to exist before this could be done for large amounts of data within reasonable amounts of time.

Beyond Backpropagation

Backpropagation spurred the development of supervised learning, leading to numerous advances in the 2010s. But AI researchers soon started noticing its flaws, and they started looming large as AI models got larger and more complex. Copious amounts of data were required to override these flaws, and a lot of tinkering with the model's parameters.

In fact, Geoffrey Hinton went so far as to say that backpropagation had outlived its usefulness and should be discarded. Quoting Max Planck, he stated, "Science progresses one funeral at a time. The future depends on some graduate student who is deeply suspicious of everything I have said."[18] On the surface, it was a surprising pronouncement from someone looking back at his life's work, but it aligns with his quest to find genuine intelligence. Another luminary of artificial intelligence, Yoshua Bengio, agreed with him that using the brain as a model for intelligence, it's unlikely that supervised learning will get us to real intelligence in artificial networks.[19]

This view was not new: as far back as 1991, cognitive scientists had proposed alternatives.[20] but the computational plausibility of these approaches remained hard to find. However, finding computationally feasible methods to replace backpropagation proved to be a challenge. Ongoing research has pursued various alternatives to backpropagation, but, thus far, these proposals remain just that—potential alternatives—without having successfully dethroned backpropagation as the dominant method.

Common Deep Learning Models

As the field of artificial intelligence continues to grow and evolve, different neural network models have risen to prominence for various types of data processing. In the early days of the resurgence of AI, computer vision models dominated the scene, powered by a structure called convolutional neural networks (CNN). Then, interest in applying neural networks to sequential data like text and speech led to a different type of model called recurrent neural networks gaining traction.

Convolutional Neural Networks (CNN)

Convolution refers to a particular mathematical operation used within some layers of a neural network, a way of compressing information but retaining its characteristics. CNNs were developed by researchers who are today's AI luminaries, such as Yann LeCun and Yoshua Bengio, who we saw in the previous chapter. The deep learning architecture they developed based on convolutions significantly improved computer vision. Models such as DanNet, AlexNet, ResNet, and Yolo were all based on a type of neural network called convolutional neural networks (CNNs).

Recurrent Neural Networks (RNN)

CNNs worked well for certain types of data, images, and videos, which could be considered individual images. But for sequential data, such as language, another approach was needed. With awkward names such as long short term memory (LSTM) or recurrent neural networks (RNN), these models incorporated memory within themselves. They could remember what came before. These models revolutionized machine-based translation across languages. To understand what RNNs do, think about reading a book. You will come across a sentence that does not make sense on its own, but when you read it with the previous sentence, understanding falls into place. For example, "The dog chased the ball. It was blue." The second sentence is ambiguous on its own, but when you read it together with the previous sentence, you can understand that the "it" in the second sentence probably refers to the ball.

An RNN works similarly—it processes information sequentially, one word at a time, and remembers what it has processed before. So, when it encounters a word that depends on information from earlier in the sequence, it can use that information to make a prediction or generate an output. For example, if you were using an RNN to generate text, and you fed it the beginning of a sentence "I like to eat," the RNN could generate the next word based on what it has learned from the previous words in the sequence. If the RNN has learned "I like to eat pizza," it might predict that the next word is "pizza."

RNNs showed two problems, though. They couldn't compete against human translators in conveying nuance, and a technical problem called the "vanishing gradient," which severely limited how far back they could remember. For example, if at the beginning of a chapter the writer had written, "August is a humid month in Paris, its inhabitants empty out into the countryside." Then, several pages later, a sentence occurs, "Their respite in the more temperate regions may last a month," it won't know that "they" refers to the absconding Parisians.

These are not unsolvable problems, but the solutions that exist are computationally very expensive. Therefore, these methods soon succumbed to the power of transformers.

Transformers

Bursting into the field of AI with a 2017 paper with the deceptively simple title, "Attention Is All You Need," transformers proved to be much more efficient and could compute in parallel. They enabled the development of models like Generative Pretrained Transformer, whose most famous version is ChatGPT.

In addition to being cheaper to train and deploy, transformers are better than RNNs in generalizing new data. For example, suppose we have an RNN model that is trained to predict the recipe for a dish based on a description of the ingredients and cooking instructions. The model is trained on a dataset of Indian food recipes and achieves a high accuracy on the training set. However, when we evaluate the model on a dataset of Inuit food recipes, we find that the model's performance is much lower, indicating that it has overfit to the Indian food recipes and cannot generalize well to Inuit food recipes.

This is because the RNN model has learned to capture the statistical patterns in the Indian food dataset, such as the common ingredients and cooking techniques used in Indian cuisine, but it may not have learned the more general principles of recipe prediction that would apply to any type of cuisine. For example, the RNN model may have learned to associate certain words and phrases with specific types of Indian dishes, such as "garam masala" with Indian curries, but it may not have learned to generalize this knowledge to other types of cuisine.

On the other hand, a transformer-based model may be better suited for this task, as it can learn more general principles of recipe prediction that apply to any cuisine. The self-attention mechanism in transformers allows them to understand the relationship between any two words in the input sequence, enabling them to capture long-range dependencies and relationships between ingredients and cooking instructions that are not directly adjacent to the input sequence. This can help the transformer model to better capture the complex relationships between ingredients and cooking instructions in different types of cuisine and avoid overfitting the specific details of the Indian food recipes.

That being said, the performance of the transformer model when tested on Inuit food recipes will still depend on the similarity between the Indian and Inuit cuisines in terms of the ingredients used, cooking techniques, and other factors. If the differences between the two cuisines are too significant, the transformer model may still struggle to generalize to the Inuit food recipes without additional training. Nonetheless, transformers are generally more robust to domain-specific variations compared to RNNs and can be expected to perform better in such scenarios.

Transformers are also able to handle rare words. Because they spread their attention wider, often across the whole input sequence—a sentence, paragraph, or longer passage, they can infer the meaning of words it has never seen based on this context. A passage about air travel safety might contain this sentence, "In addition, the redundancy extends to the nogginsapper, where the current is always constant so the brakes can be manually set." It will infer from the surrounding context that the nogginsapper is an electrical system with redundancies built into the airplane, probably used for braking.

Suppose we have a transformer model that is trained to predict the sentiment of a movie review as either positive or negative. When the model is fed with a movie review such as "The acting was good but the plot was weak," it needs to understand the hierarchical structure of the sentence to correctly classify the sentiment. Specifically, it needs to recognize that the positive sentiment towards the acting and the negative sentiment toward the plot are both present in the same sentence and should be combined in some way to arrive at the overall sentiment of the review.

The transformer model can learn to represent the sentence in a hierarchical way by using multiple layers of self-attention. At the lowest layer, the model can attend to individual words in the sentence and learn their representations based on the context in which they appear. At higher layers, the model can attend to groups of words and learn representations that capture more complex syntactic and semantic relationships between the words.

For example, at the second layer, the model may attend to the phrase "the acting was good" and learn a representation that captures the positive sentiment toward the acting. Similarly, at the third layer, the model may attend to the phrase "the acting was good but the plot was weak" and learn a representation that combines the positive sentiment toward the acting and the negative sentiment toward the plot to arrive at the overall sentiment of the review.

This hierarchical approach allows the transformer model to capture complex relationships between words and phrases in a sentence and arrive at a more accurate sentiment classification. This is in contrast to simpler models such as bag-of-words or RNN models, which may struggle to capture the hierarchical structure of the sentence and the interactions between its constituent parts.

Foundational Models

ChatGPT has popularized the concept of foundational models. This term is used to describe models that draw upon vast amounts of data for training and can be adapted for use in a wide variety of downstream tasks. To grasp their significance, we can draw a parallel with the emergence of higher-level programming languages such as BASIC and C in computer programming. In the early days, programmers had to write intricate machine code to make computers perform specific tasks. It was a complex and time-consuming process that required a deep understanding of the underlying hardware. The emergence of higher-level languages recast the programming profession. These languages provided programmers with more intuitive syntax and abstraction layers, making it easier to write code and communicate instructions to computers. More people were able to enter the field of computer programming because these languages made it more accessible.

In the past (which is the 2010s), creating AI programs was a bit like learning to code from scratch. Developers spent a lot of time and effort understanding complex details of the hardware the code would run on, and built the AI models layer by layer. Now, foundational models like ChatGPT serve as "magical cookbooks." These are not literal cookbooks but models that have already learned a lot from vast amounts of data. Instead of grappling with the intricacies of training models from scratch, multitudes of AI experts can now leverage these foundational models trained on massive datasets. By using foundational models like ChatGPT, AI practitioners can focus their efforts on specific downstream tasks without getting entangled in the intricacies of low-level model training.

The effect of these models reverberates beyond the specialty of AI development. Coders now get snippets of code they can modify, greatly accelerating coding projects. Formal business letters can be constructed with ease, and quick research becomes easier than a Google search. The applications of these models continue to unfold.

Handle with Caution

If these foundational models were a physical product, their warning labels would come with stark warnings. Language models, especially, should be used with great caution. These models display their amazing creations from one simple instruction, which is to generate the next most likely word in the sequence based on what has come before. Note that the model is not looking up examples of text with which it was trained. Nor is it connecting to the internet to look up anything. Based on the many words in the sequence fed to it, it generates several lists of words, each with the probability of being the most suitable next word. Then, it chooses the one after that, using the same process. While it is mind-boggling to think that a simple rule like that can result in so much complexity, that's all there is to it.

However, it also leads to behavior that may be unpredictable and potentially dangerous. The intricacies lying beneath the surface of these models render them both awe-inspiring and precarious. Comparable to a double-edged sword, they possess immense power to transform mere prompts into eloquent passages, all while harboring the potential to veer off into uncharted territories of misinformation or harm.

Consider a scenario in which a magazine has seen its revenue drop due to plunging subscriptions and declining ad sales. Its publisher is bought out by a private equity firm, which fires most of the writing staff and editors. Then, it hires someone to generate articles using a Large Language Model (LLM) like ChatGPT.

That person programs ChatGPT by feeding it specific information and asking it to create a story. For example, they might provide some facts about how removing trees in cities can raise the temperature and then instruct ChatGPT to craft an article on that topic. The result is impressive: the output is free from spelling mistakes and grammatically flawless. However, ChatGPT's instruction to generate the most likely next word means it has no problem making up facts. Especially the longer the article gets, which is a peculiarity of LLMs. It lacks human judgment and context, so the remaining magazine staff should now be obligated to verify and interpret the information the LLM generates. They need to ensure that the content aligns with truth and adheres to the ethical standards of journalism.

A cautionary, real-life incident occurred when an experienced lawyer in New York used ChatGPT to create a legal brief on behalf of a client suing an airline for negligence.[21] ChatGPT gave him a brief with the correct language and format, and it cited cases with the names of real judges but completely fabricated citations and case decisions. Unfortunately, the lawyer filed the brief, and the outcome made him and his firm internationally famous. The judge dismissed the suit, and the duped lawyer had to apologize and pay a fine.

Reprieve from this drawback is imminent. New architectures that do not depend on this type of word generation will soon overshadow ChatGPT. In the meantime, output should be treated with caution and a great deal of skepticism. Human supervision is still necessary.

Vision and Sound

ChatGPT is built on a foundational model for text. Other foundational models like DALL-E 2 and Midjourney are trained on vast amounts of visual data and combined with language models. Visual models create an abstract representation of an image, and a language model creates a representation of the textual description of that image. These representations then undergo a process of cohesive fusion or association. Thus, words and their meanings become correlated with specific abstract representations of images.[22]

One can imagine similar multi-modality foundation models that associate sound and text. Imagine being able to say, "Write a movement in C minor with violins and oboes," and the output is a composition on sheet music or actual music fit for a concert hall performance. Such a model does not exist, but it is a matter of time before it does.

What does exist today is a model that can search for sounds in a piece of audio.[23] By merely describing the desired sound, one can prompt the model to scour the entire audio file for instances of that particular sound. Researchers are using it to identify and locate birds and other species from sounds captured in the Amazon rainforest by microphones. For conservationists, this is a boon to surveying the biodiversity health of vast areas.

Additionally, there are now models designed to mimic human voices. While text-to-speech technology isn't novel and has gained widespread use with virtual assistants like Siri and Alexa, a significant leap was made with Microsoft's model that could either replicate any voice or produce synthetic voices indistinguishable from native speakers. The simulations are so good that they can deceive most people. What's more remarkable is that these models are scalable; a few seconds of someone's voice is usually sufficient to create a representation in a neural network. Then, the model can be used to generate any speech in that person's voice.

Explainable AI

While neural network models have unquestionably had immense success, there are some criticisms. One of the major drawbacks of AI models as we currently structure them is the mystery around how they achieve their results. Users of AI models know the input and the output, but what happens in the middle of the calculations and transformations that turn the input into the output are only vaguely known. It is a black box; we cannot peer into it to see which wheels and gears are turning.

Well, that is not entirely accurate. We can observe the connections between neurons, but we still lack a clear understanding of how to interpret the weights associated with those connections in a way that provides us with intuitive insights into the model's functioning. Take for instance, the ability of a model to identify a cat in an image it has never encountered before. How does it accomplish this feat? What knowledge has it acquired about cats that enables such recognition? Is it based on the specific arrangement of their eyes, nose, and mouth? Or is it the rough texture of their fur, pixel by pixel, that sets them apart?

There are situations where it becomes crucial to understand how and which factors in the input lead to the model's output. Consider the task of predicting famine in a particular region using a neural network model. It may yield highly accurate results, but if the objective of applying such models is to intervene in the region and prevent famine altogether, then researchers and policymakers need to know how to intervene to head off famine. Are there specific factors such as rainfall patterns or soil composition that are driving the occurrence of famine? Traditional models possess an advantage in that they can guide researchers in identifying the underlying causes of famine.

Of course, AI researchers have not been idle. Methods are being developed to probe these black boxes and understand how they make their decisions and how to interpret the features they find important. In fact, further development of AI models may depend on some insight being offered into their inner workings. While pondering the responsible development of AI, governments or regulatory bodies might require developers to explain what led their models to their conclusions.

Final Thoughts

The evolution of AI will continue to be mirrored by the language used to describe it. Words like transformers did not exist in the lexicon 10 years ago, but now have taken on a new meaning in addition to their original meanings. Many of the terms detailed in this chapter come from 20th century psychology and computer science. We can anticipate a surge of neologisms entering the lexicon with the emergence of technologies like explainable AI, federated learning, and quantum machine learning. The new terms will probably be coined by researchers introducing them in their publications or perhaps even by corporations seeking to market and popularize their AI-driven products and services. For the AI practitioner, an understanding of this specialized language is a useful first step in making a mark in the field.

References

1. Samuel, A. L. Some studies in machine learning using the game of checkers. *IBM Journal of Research and Development* **44,** 206–226 (2000).
2. Hilgard, E. R. & Bower, G. H. *Theories of Learning.* (Prentice-Hall, 1975).
3. Thorndike, E. L. *The Psychology of Learning.* (Teachers College, Columbia University, 1913).
4. LaBar, K. S. & Cabeza, R. Cognitive neuroscience of emotional memory. *Nat Rev Neurosci* **7,** 54–64 (2006).
5. Schwartz, B. L. & Metcalfe, J. Tip-of-the-tongue (TOT) states: Retrieval, behavior, and experience. *Memory & Cognition* **39,** 737–749 (2011).
6. Kirkpatrick, J., Pascanu, R., Rabinowitz, N., Veness, J., Desjardins, G., Rusu, A. A., Milan, K., Quan, J., Ramalho, T. & Grabska-Barwinska, A. Overcoming catastrophic forgetting in neural networks. *Proceedings of the National Academy of Sciences* **114,** 3521–3526 (2017).
7. Bjork, D. W. B.F. *Skinner: A Life.* (Basic Books, 2009).
8. Watson, J. B. *Behaviorism.* (W.W. Norton, 1924).
9. Silver, D., Huang, A., Maddison, C. J., Guez, A., Sifre, L., Van Den Driessche, G., Schrittwieser, J., Antonoglou, I., Panneershelvam, V. & Lanctot, M. Mastering the game

of Go with deep neural networks and tree search. *Nature* **529,** 484–489 (2016).

10. Moyer, C. How Google's AlphaGo bear a Go world champion. *The Atlantic* (2016). at https://www.theatlantic.com/technology/archive/2016/03/the-invisible-opponent/475611/
11. Lent, R., Azevedo, F. A. C., Andrade-Moraes, C. H. & Pinto, A. V. O. How many neurons do you have? Some dogmas of quantitative neuroscience under revision. *European Journal of Neuroscience* **35,** 1–9 (2012).
12. Finger, S. *Origins of Neuroscience: A History of Explorations into Brain Function.* (Oxford University Press, 2001).
13. Swanson, L. W., Newman, E., Araque, A. & Dubinsky, J. M. *The Beautiful Brain: The Drawings of Santiago Ramon y Cajal.* (Abrams Books, 2017).
14. Ehrlich, B. *The Brain in Search of Itself: Santiago Ramón Y Cajal and the Story of the Neuron.* (Farrar, Straus and Giroux, 2022).
15. Minsky, M. & Papert, S. A. *Perceptrons: An Introduction to Computational Geometry.* (The MIT Press, 2017). doi:10.7551/mitpress/11301.001.0001
16. Rumelhart, D. E., Hinton, G. E. & Williams, R. J. Learning representations by back-propagating errors. *Nature* **323,** 533–536 (1986).
17. Chauvin, Y. & Rumelhart, D. E. *Backpropagation: Theory, Architectures, and Applications.* (Taylor & Francis, 2013).
18. LeVine, S. Artificial intelligence pioneer says we need to start over. *Axios* (2017). at https://www.axios.com/2017/12/15/artificial-intelligence-pioneer-says-we-need-to-start-over-1513305524
19. Bengio, Y., Lee, D.-H., Bornschein, J., Mesnard, T. & Lin, Z. Towards Biologically Plausible Deep Learning. Preprint at https://doi.org/10.48550/arXiv.1502.04156 (2016)

20. Mazzoni, P., Andersen, R. A. & Jordan, M. I. A more biologically plausible learning rule for neural networks. *Proceedings of the National Academy of Sciences* **88,** 4433–4437 (1991).
21. Mulvaney, E. Judge sanctions lawyers who filed a fake ChatGPT legal research. *Wall Street Journal* (2023). at https://www.wsj.com/articles/judge-sanctions-lawyers-who-filed-fake-chatgpt-legal-research-9ebad8f9
22. Radford, A., Kim, J. W., Hallacy, C., Ramesh, A., Goh, G., Agarwal, S., Sastry, G., Askell, A., Mishkin, P. & Clark, J. Learning transferable visual models from natural language supervision. In *International conference on machine learning* 8748–8763 (PMLR, 2021).
23. Wu, Y., Chen, K., Zhang, T., Hui, Y., Berg-Kirkpatrick, T. & Dubnov, S. Large-scale contrastive language-audio pretraining with feature fusion and keyword-to-caption augmentation. In *ICASSP 2023-2023 IEEE International Conference on Acoustics, Speech and Signal Processing (ICASSP)* 1–5 (IEEE, 2023).

3

AI for Good

"Life's most persistent and urgent question is, 'What are you doing for others?'"

– Martin Luther King Jr.

"Technology is nothing. What's important is that you have faith in people, that they're basically good and smart, and if you give them tools, they'll do wonderful things with them."

– Steve Jobs

WHEN TECH LUMINARIES such as Bill Gates and Elon Musk were warning against the dangers of AI,[1] there was a reaction against this dismal view from several quarters, and the phrase AI for Good was born.

AI for Good is when artificial intelligence is harnessed to improve livelihoods and the environment. It can mean using AI to solve global problems, such as climate change, food security, and health. AI for Good exists not just as a concept; it is a movement driven by the conviction that AI has the potential to significantly benefit humanity.

We count fire and electricity as major milestones in the evolution of human society. It is tempting to believe that AI may soon join their ranks.[2] Just as fire changed the course of our physical and cultural evolution by giving us warmth and a means to cook our food, and electricity propelled us into the information age and a new era of productivity, AI has the power to revolutionize our future in ways we can only imagine right now. It is a force that can propel us toward positive change, but also one that requires careful consideration and responsible action.

As we will see in later chapters, AI can also be misused and manipulated for selfish gains. The risks are real, and the ethical concerns are valid. We will look at efforts to imbue AI development with ethical considerations, fairness, transparency, and accountability.

In this chapter, we will see how AI for Good manifests in fields such as medicine, farming and food security, water security, environmental sustainability, and biodiversity.

Responding to Natural Disasters

Climate change has wrought havoc on weather patterns that human society has already adapted to over millennia. As the Earth struggles to adapt to greater concentrations of greenhouse gases, extreme weather events are becoming more common. In AI, we may soon have an invaluable tool to help us address these challenges. Using AI, we can quickly analyze data from various sources, such as satellite imagery, weather reports, social media posts, and ground sensors to gather real-time information on the disaster and its impact. This wealth of information can be used to prepare for potential disasters, coordinate rescue efforts, and ultimately save lives.

Let's consider two recent scenarios where devastating natural disasters spurred development of AI systems to reduce the scale of future, similar disasters.

Floods

Think of Sameer, a farmer in Sindh province in Pakistan in 2022. Sameer relied on the Indus River to grow his crops of mangoes, sugarcane, rice, and wheat. But he knew the river could easily take away his livelihood, as it had done in the past. His neighbors and relatives warned him that flooding could destroy his house and all his possessions—his trunk where his wife and three school-age children keep their clothes, his bed with the woven hammock, and a few cooking utensils. Without crop insurance, which he couldn't afford even if it was available, he can only hope that this day will never come. But if it does, he puts his faith in his special deity, offering prayers every morning, knowing that fate is out of his control.

Unfortunately, he was likely one of the millions that were affected by the floods caused by the monsoon rains in Pakistan in 2022. The floods destroyed 300,000 houses, killed 1,200 people and displaced 3.1 million more. It was not a once in a lifetime disaster. Sadly, it was not even an unpredictable disaster. In 2010, floods in the same region took almost 2,000 lives, and displaced 6 million people.[3]

Once a flood has already devastated a population, an army of nongovernment organizations (NGOs) springs into action to succor the victims. This is where AI can play a role as well. Aerial images and satellite images can help NGOs identify with precision which areas need help.

Can't we predict these tragedies before they happen? Scientists have models that can predict floods, sometimes with a lead time of days. Coarser predictions can be available weeks or months before a potential event. There is data flowing in from satellites, weather balloons, and whole arrays of sensors. However, deploying AI to sift through all this data and provide warnings of floods is still in its infancy. Pakistan alone has around 7,000 glaciers whose meltwater adds to the rivers and streams. While it hasn't been conclusively proven, glacial lake outburst floods have been major factors in past floods, such as the Batura Glacier flood in 2010 and the Lugge Tsho glacial lake floods in Bhutan in 1994.

The World Meteorological Organization is coordinating efforts by dozens of companies and research organizations to create the infrastructure and technology for early warning systems.[4] However, putting in place global systems requires a lot of time and energy, and more nimble, local solutions may help pave the way, such as one championed by an organization called SEEDS in India.

SEEDS

The emergence of early warning and preparedness systems as a civic necessity may be happening in India. Anshu Sharma and Manu Gupta were witnesses to many floods in the town where they grew up, New Delhi. In college, Anshu became convinced that customized advice to individual households could mitigate the effects of floods in urban areas. A grassroots approach combined with AI technology is the synthesis that they are forging through their organization, SEEDS, which stands for Sustainable Environment and Ecological Development Society.

The Microsoft AI for Good Research Lab developed an AI model that used high-resolution images from Airbus to map two regions. Local experts, with SEEDS guidance, created a flood risk scale at a city-block resolution. The 5-point scale indicated propensity to flood, with 5 being the worst. The experts determined this number by combining output from the AI model with local knowledge of building types, topology, and other relevant data.[5]

A real-world validation of this process occurred during cyclones in two other regions in India, where the model's accuracy proved to be satisfactory. SEEDS now wants to expand their system to six more cities in India, promising to provide valuable guidance to city officials in areas likely to be most impacted.

The goal of this initiative is ambitious: to revolutionize risk management with a ground-breaking early warning system. City officials and relief organizations can use a dynamic dashboard that continually updates the risk profile of each block. This would allow for the establishment of resource depots strategically positioned to facilitate flood rescue and survival, with sufficient notice to set up response centers in a matter of days. With just a few hours' notice, areas at risk can be evacuated, mitigating potential harm and safeguarding vulnerable communities. This proactive approach to disaster management holds great promise in enhancing preparedness and response capabilities, making a positive impact on the lives of those in flood-prone regions of India.

SEEDS did not materialize overnight. Its inception dates back to 1994, and it took the dedication of two exceptionally bright and driven people to make this dream a reality. Their enthusiasm and determination were the driving force behind the success of this AI project, just as they are for any other successful endeavor.

Earthquake in Syria and Turkey

Earthquakes are not directly affected by climate change, but they are a common occurrence, and the same tools used after floods and fires can be used to save lives in their aftermath. The 2023 earthquake that struck Syria and Turkey was one of the most devastating natural disasters in recent memory. It struck without warning, leaving unimaginable despair in its aftermath. News broadcasts showed apartment buildings reduced to rubble, streets filled with debris, and vibrant communities turned into scenes of devastation.

The human toll of the earthquake was immense. The number of casualties climbed exponentially as search and rescue teams worked tirelessly to reach those trapped under the rubble. Initial estimates of 3,000 dead were soon overtaken by 4,000, then 5,000. The eventual count of dead people came to 56,000. One and a half million more were left homeless and without access to basic necessities.[6] In neighboring Syria, where the earthquake had effects just as devastating, the full impact was unclear due to the ongoing civil war.

In the face of such widespread devastation, the government of Turkey found itself at the epicenter of a torrent of public outrage. It was clear that their decisions, driven by political expediency, had contributed significantly to this devastating social calamity. Despite the region's known history of seismic activity, building regulations had been callously suspended, serving as a means to solidify the ruling party's support base, but ultimately resulting in the loss of countless lives.

The significance of political and social attitudes in disaster mitigation cannot be overstated.[7] Take Japan, for example, where despite 15,000 lives lost in the 2011 earthquake and tsunami, buildings remained intact, and countless lives were saved due to effective measures. AI can certainly provide better tools for response, but it is the leaders with the will to utilize them who truly make a difference.

In the aftermath of the disaster in Turkey numerous organizations stepped up to provide support to relief forces on the ground. Satellite image providers such as Planet Labs and Maxar supplied critical images to the Turkish disaster response agency, AFAD. Technology companies offered AI-supported analyses of these images, which helped relief organizations quickly prioritize evacuations, search and rescue, and relief services. These collaborative efforts highlight the importance of coming together in times of crisis to aid those in need and mitigate the devastating effects of disasters.

Role of AI in Disaster Resiliency

While AI still hasn't cracked the complex challenge of disaster prediction, it can be used to direct post-disaster responses. That these disasters are becoming more frequent is beyond doubt. It is not just a by-product of greater information connectivity around the world; the world's climate is changing, and weather patterns are in a state of flux.

AI systems enable fast, highly scalable processing of vast amounts of data. They can adapt to new types of data with perceptual and cognitive skills equal to hundreds or thousands of human analysts.

However, to truly address these impacts of climate change, we need more than just rapid response. We need comprehensive strategies that combine the strengths of AI with policy decisions and collective climate action. We need leaders in politics and science, private companies and social organizations, who prioritize climate action and make smart policy decisions. And to help these leaders understand the risks and vulnerabilities associated with climate events, we need accurate data.

Data is a versatile concept encompassing various forms of information. For instance, an individual's personal account of witnessing the river's gradual ascent constitutes data. It can be used by individuals to take preventive or exploitative action. Similarly, a meticulous record of water levels collected at regular intervals by a monitoring station also falls within the realm of data. This data can be employed by automated systems to produce timely notifications. Artificial intelligence systems heavily rely on the latter data type, requiring substantial quantities to operate effectively.

By synthesizing factors like temperature, precipitation, and rising sea levels while simultaneously evaluating vulnerabilities within susceptible regions and populations, AIs can be developed to provide early warnings of looming natural disasters and help direct post-disaster efforts. As of 2023, satellite images were the workhorse of AI applications dealing with climate change issues. In the immediate period following a catastrophic event, satellites are tasked to monitor the devastated area. They pinpoint where devastation has occurred, and they can be correlated with population maps to identify where people may have been caught by the disaster.

Presently, we possess the capability, owing to comprehensive population maps provided by WorldPop and the IHME-Microsoft-Planet alliance,[8] to ascertain the locations of people, not at the individual level, but at the aggregate at a very fine geographical level: within 100m by 100m from WorldPop, and even finer through Microsoft. The advantage of these maps powered by satellite images and AI is that they can identify changes in populations almost immediately, compared to decennial country censuses.

Data for Environment Monitoring

There is a wealth of data available that AI researchers are already using. Let us look at some of the types of data that are useful for disaster prediction and management.

Remote Sensing Technologies Once the domain of militaries and governments, satellites have become much more widely available. They are invaluable for civilians working in communications, agriculture, humanitarian organizations, and the list goes on. Landcover mapping is an indispensable tool in many of these endeavors. From the lush forests of the Amazon rainforest to the barren expanses of the Sahara Desert, it lays bare the secrets of the planet's surface in meticulous detail.

The pioneering Landsat satellites program, which originated in 1972, played a pivotal role in granting access to this invaluable data. Run jointly by NASA and the US Geological Survey, its ninth iteration, Landsat 9, continues utilizing increasingly powerful sensors across the visible and infrared spectrum to serve a multitude of purposes. Landsat images were even used to find the possible final resting place of Amelia Earhart.[9]

Free images are also available from providers such as Sentinel II, operated by the European Space Agency. A number of commercial providers also may offer lower resolution versions of their products for free or low cost, such as Planet Labs and Maxar. In 2020, the Norwegian government made a momentous investment in using satellite imagery to combat deforestation. In a deal with Planet and Airbus, it made satellite data for all the world's major forested regions available to everyone.[10]

The sheer amount of data amassed by these images is staggering. They offer an unprecedented scale of information, depicting various vegetation types, soil and rock compositions, water bodies, and human settlements across the globe. Scientists then construct models for a multitude of applications—vegetation cover, flood risk, drought risk, weather, and many more uses are being found every year.

For flooding, the satellite images can be analyzed to determine where bodies of water may overflow. Correlating this with topographic data and precipitation from radar satellites can help with early prediction. Monitoring glaciers and predicting the possibility of a lake outburst flood are important in certain locations where they may threaten human settlements, as happened in Pakistan.[11]

NASA has several satellites in use for monitoring destructive wildfires.[12] Given the wide availability of satellite data and the spread of AI expertise across the globe, it is no surprise that several startups are now combining all these elements to help tame wildfires. Their ultimate goal, tantalizingly within reach, is to be able to predict where they will emerge.[13] Satellites can be among the first to detect fires starting in remote regions, using data from a range of sensors that include visible light, thermal and infra-red, moisture content of the ground, and radar data about weather conditions.

Satellites have made tremendous progress in their ability to capture images of vast areas of the Earth's surface, with resolutions ranging from several kilometers down to less than a meter. This wide coverage proves sufficient for many applications, but it can sometimes be costly to obtain high-resolution satellite data. Remote sensing can also be done by aircraft, including drones. They cannot cover as much ground as satellites, but they excel in providing incredibly detailed imagery at resolutions even finer than a centimeter. This kind of remote sensing is valuable for tasks such as detailed mapping of regions and intense disaster rescue and recovery operations.

Atmospheric Data For predicting floods and wildfires, weather data stands out as a crucial resource. Scientists' models can ingest a rich variety and amount of information, such as temperature, precipitation, wind speed, and humidity. These models can then paint a picture, metaphorically, of rivers overflowing because of unusual rainfall, and they can provide these warnings in good time to mitigate the impacts of flooding. The weather data landscape is vast, encompassing thousands of monitoring stations scattered across the globe. There are also space-borne radar satellites that provide incredibly accurate readings every hour or less.

Weather prediction has traditionally been based on modeling physics. But some researchers, seeing what AI does with discovering novel features in large data, decided to let go of their physics equations and feed only observations into an AI model. The resulting model, called

Pangu-Weather, is reported to be 10,000 times faster than traditional models, with similar levels of accuracy.[14,15]

Seismic Data The collection of seismic data is a global endeavor, with thousands of monitoring stations strategically positioned across the Earth. These stations capture seismic waves and record ground motion during earthquakes and other seismic events. Where these stations do not exist, techniques similar to ultrasound imaging can create maps of the techtonic volumes under the ground. With this data, scientists can create detailed maps of the subsurface structure, including beneath the ocean floor. This information is highly financed by oil companies, but is incredibly valuable for geological research as well.

Seismic data is more fodder for AI models.[16] Since these models excel at finding patterns and correlating them with specific outcomes, there is reason to hope they can unlock the potential to foresee earthquakes and other seismic phenomena with unprecedented accuracy and early enough to prevent loss of life.

Population Data For disaster planning and response, knowledge of where people live is of paramount importance. The intricate mosaic of human settlements, often changing faster than a decennial census, necessitates more up-to-date representations. If detailed estimates are available from local governments, they should be used, but disasters often happen in parts of the world that do not have this luxury. One approach that AI researchers have taken is using satellite images to look at rooftops.

Serving as proxies for people living there, we can get a fairly good estimate of the distribution of people in most regions of the world.

WorldPop has developed and maintained an exceptional dataset of populations around the world. Their dataset goes beyond mere population counts and incorporates additional demographic characteristics such as age, gender, and socio-economic status. To achieve this level of detail, WorldPop combines satellite imagery with information obtained from household surveys and census data.

The richness of this dataset has made it invaluable for a wide array of applications. Used in planning vaccination campaigns, it enables authorities to target specific population groups effectively. In disaster response efforts, the dataset aids in understanding population distributions, helping efficient allocation of resources. It also proves useful in tracking the spread of infectious diseases, enabling researchers and healthcare professionals to monitor population movements and identify areas at higher risk.

Urban planners and policymakers also gain insights into the evolving socio-economic characteristics of urban areas. This information can inform the development of sustainable urban infrastructure and support evidence-based decision-making. Remarkably, this data is available to anyone who wishes to use it, making it an invaluable resource for researchers and policymakers. Recognizing the fundamental utility of these types of datasets, similar efforts underway by the Microsoft AI for Good Lab, Planet, and the Institute for Health Metrics Evaluation will refine these maps at even finer resolutions.

Cellular and Emergency Call Data If police can subpoena mobile phone records to pinpoint locations of criminals, then surely emergency responders can immediately locate a person in distress by accessing their mobile records or cell-tower pings. Sifting for signals through this data is a large, complex challenge. By analyzing changes in call and text message patterns, as well as changes in mobile network activity, responders can quickly identify where people are, which can inform decisions about where to allocate resources. They can also identify where power outages have occurred or can be used to send alerts.[17]

Like so many systems at this moment in AIs history, we are merely at the beginning of training AIs to take advantage of this wealth of data. The implementation of such a system is not without its challenges. Privacy concerns are paramount in this discussion, as the use of mobile phone records for emergency response purposes would necessitate a delicate balance between public safety and individual privacy rights. We explore this further in Chapter 8.

News and Social Media The potential of social media and crowdsourced data in helping in the recovery and rescue periods after a disaster is tantalizing. One can imagine language models tirelessly scouring social media posts and images during a flood event, working in concert with other AI models in putting together real-time information about local conditions and risks.

However, this promise has yet to be borne out. For instance, during the 2018 Kerala floods in India, messaging platforms such as WhatsApp helped communications among rescuers and people needing help. However, disinformation was rife during the floods, with false information and doctored images causing confusion among affected populations. Disinformation, textual or doctored images, is just one of the challenges an AI system would face with social media.[18]

Nonetheless, systems like FloodTags have claimed some successes in the news media monitoring approach to data collection. While not as up-to-the-minute as social media, news media may potentially offer early awareness of major events around the world. The scope of this challenge cannot be overstated, with millions of items generated in hundreds of languages every day. In other words, we need intelligent, automated systems to process and sift through this data.

Food and Water Security

Climate change exacerbates floods, tornadoes, and wildfires. The change in rainfall patterns and increasing temperatures will likely disrupt food practices. Poorer, smallholder farmers will be most vulnerable since adaptation may be too costly. This disruption will be felt in the oceans, too. Depleted fisheries will suffer from warmer temperatures.

In 2022, a visit to the farmlands in central Kenya or pastures near Mount Kilimanjaro would have shown the carnage that drought creates. My family and I were visiting Amboseli National Park, once home to a river-fed lake attracting bird watchers from all over the world. Instead, we flew over a dry river valley and landed to the stench of rotting animal carcasses that littered the park. On a typical safari drive, one views herds of gazelle and wildebeest (or whatever animals you want to highlight) gathered together on alert for predators. During our trip, there were no herds, just isolated animals, each searching for any patch of vegetation to stave off starvation. They didn't need to worry about lions or hyenas as they lay nearby, with stomachs distended from having eaten too much. Wildebeest were falling over before our very eyes, starving and too weak to get up. There were simply too many dead for the predators to eat any more.

These wild animals were competing with domesticated herds of cattle and goats, driven by people who lived near the park, desperate to find sustenance for the livestock that were their livelihood.

In Northern Kenya, a heart-wrenching reality unfolds as families struggle to survive in the face of famine and drought. These families are not thinking about next week or next month—they are focused on surviving the next hour. For them, every day is a battle for survival.

In many cases, starvation is the most extreme manifestation of food insecurity, where individuals are unable to access sufficient food to meet their basic nutritional needs. However, food insecurity encompasses a wider spectrum of challenges, ranging from inadequate access to nutrition, to issues of food safety. The Food and Agriculture Organization's FAO's definition is: "[the] situation when people lack secure access to sufficient amounts of safe and nutritious food for normal growth and development and an active and healthy life." Severe food insecurity is most pronounced in Africa and South America, but it is present to varying degrees worldwide. In sub-Saharan Africa, for example, the percentage of people experiencing severe food insecurity passed 20% in 2016. Currently, over 1 billion people globally are facing severe food insecurity.

Water insecurity goes hand in hand with food insecurity. Water is not only crucial for sustaining living bodies but also plays a vital role in growing food, hygiene, and sanitation.

The lack of access to water is a harsh reality for millions of people around the world. A visit to any smallholder farmer in the Global South will reveal the precarious edge on which they exist. Take a smallholder like Ayanda in South Africa, who relies on rainfed agriculture to support her family and make a living. For Ayanda, each season is a gamble as she struggles to predict when and how much rain will fall. The failure of her crops not only reduces her income but also deprives her family of food and nutrition. To make matters worse in low rainfall periods, Ayanda has to walk long distances to fetch water from a contaminated river, putting herself and her family at risk of waterborne diseases. And the lack of proper sanitation facilities only adds to their misery. Soap and water are not guaranteed.[19] Ayanda's situation illustrates the state of at least a quarter of the world's total population.

While this may seem like a distant problem to many of us, it is a reality for millions of people around the world. And the worst part? It is a predictable crisis.

Food and water insecurity are age-old problems. But we've reached a point in human history where it is within our grasp to eradicate those ancient afflictions. Several non-governmental organizations have been working in this direction for years. They have supplemented governments in trying to bring people relief.

For a long time, only a few organizations had the resources—subject matter experts, relevant data, etc.—to make forecasts of food insecurity, famine, and drought. One significant aspect of the AI revolution is the democratization of access to data. Organizations, individuals, and even those with limited resources are able to contribute to disaster prediction and response efforts. Organizations with expertise in agriculture and humanitarian aid can now take advantage of AI to bolster their impact.

Emerging Uses of AI to Combat Food and Water Insecurity

To improve agricultural yields, AI can help with landcover mapping, crop and disease monitoring, precision agriculture, energy and irrigation, markets and supply chains, and training and education.

Landcover Mapping Satellite images are being used to analyze landcover and soil types. Farmers and scientists can gain insights about the health and productivity of the land and identify areas that are suitable for agricultural activities. The satellite company Planet maps the Earth every single day. Similar companies are enabling a continuous record of the Earth at an unprecedented detail and scale. Anyone with the available GPUs can monitor change in land use patterns over time. This information can be used to optimize farming practices, such as identifying areas with low vegetation cover that may require additional soil enrichment or identifying areas with high vegetation cover that may indicate potential pest infestations or invasive species.

NASA Harvest undertook a comprehensive land-cover survey of Ukraine in 2022. By analyzing daily satellite images from Planet, they aimed to assess the potential impact of the Russian invasion on Ukraine's agricultural sector and the global food supply. Their evaluation showed that Ukraine's agriculture remained remarkably resilient.[20] This insight provided reassurance amid concerns about under-harvested wheat and other crops that could have contributed to worldwide food shortages. "Satellite data enables us to provide rapid agricultural assessments that are critical for markets and food security," observes Inbal Becker-Reshef, the program director of NASA Harvest. Her statement is poised to be validated repeatedly in the coming years.

Crop Monitoring Computer vision technologies similar to those used to identify pictures of cats or tagging family members in social media photos can also be used on photos or videos of crops and fields. The AI algorithms can detect signs of pest damage, disease, or nutrient deficiencies.[21] This can enable farmers to identify and address these issues early, preventing potential crop losses and increasing overall yield.

Precision Agriculture AI can provide valuable insights for precision agriculture, using data called edge devices. These devices operate far from the cloud, on the edge, so to speak. They are devices such as moisture and pH sensors, and animal collars. For livestock, wearable devices can track their movements and monitor vital signs and nutritional needs. In agronomy, AI can analyze weather data to help farmers make informed decisions about irrigation schedules, fertilizer application, and pest management.

FarmBeats, developed at Microsoft, combines affordable agricultural tech to provide exactly these insights. With water destined to become a scarce resource in the near future, accurate and targeted water delivery to crops becomes crucial for the smallholders and farmers to sustain themselves.

Energy and Irrigation Optimizing water use for agriculture is an essential part of the strategy for heading off food insecurity. The Nature Conservancy (TNC) is actively leveraging satellite maps (Figure 3.1) to identify and delineate smallholder farms situated on the Northern slopes of Mount Kenya. This mapping effort will provide a detailed understanding of irrigation practices in the region and identify areas where water usage can be optimized. Combined with a knowledge of the spatial distribution of farms, TNC can develop targeted strategies to enhance irrigation efficiency and reduce water waste. This initiative is part of TNC's Foodscapes program, which takes a holistic approach to improving food production in vulnerable areas around the world.

In addition to satellite mapping, AI also plays a more direct role in optimizing irrigation practices. With soil moisture data, crop water requirements, and several other factors, AI systems can optimize irrigation schedules. This data-driven approach ensures that crops receive the appropriate amount of water at the right time, avoiding both under- and over-irrigation.

Figure 3.1 Satellite image of farms in The Nature Conservancy's CHEF region

Source: Bing Maps

By optimizing irrigation, farmers can not only conserve water but also reduce energy costs associated with pumping and distribution.

Central to the success of food security strategies is the availability of affordable and sustainable energy sources. To address this need, self-contained mini-grid systems are being implemented, leveraging renewable energy resources like wind and solar power. In many regions of the Global South, power infrastructure remains greatly under-developed. The main electricity grid cannot make it to all populations; hence, self-contained grids generating power in the range of just a few kilowatts to 10 megawatts[22] are being built by several NGOs.[23,24]

One fascinating approach to providing electricity in underserved areas was initiated at Virunga National Park in the Democratic Republic of the Congo. This area is under threat from several militia groups fighting a civil war, and much of the local population is caught in a poverty trap. Access to electricity is seen as a key means to escape this trap and create a peace economy that will weaken the war economy that has taken hold. They set up a utility called Virunga Energies that generates hydroelectricity and distributes it via a smart grid to households and local micro-businesses (Figure 3.2).

Figure 3.2 Photo of Matebe hydroelectric powerplant that was built and is run by Virunga Energies, the utility subsidiary of Virunga National Park. It is located on the edge of the Virunga National Park and feeds from the river networks that flow from the Mountain Gorilla Sector. The mountains of the Mountain Gorilla Sector can be seen in the background.

Credit: Virunga Energies

The innovation comes in the form of electricity loans. When a micro-business is approved for a loan, a balance is created in their utility account. Payments they make will be used to repay the loan. AI can play a partial role as well. Just as large power companies can use AI for efficient operations, a company like Virunga Energies can use machine learning in the same way banks use credit risk to determine loans.

Global Renewables Watch Renewable energy and food security do not always go hand in hand. Tension arises from the fact that the same land can be used generating renewable energy or cultivating crops to support food production, or for other uses entirely. This presents a complex challenge as governments need to carefully navigate and balance the utilization of land resources for both renewable energy and agriculture.

Recognizing this, The Nature Conservancy[25] worked with Microsoft's AI for Good Research Lab and Planet Labs to gather data to determine the extent of this trade-off. Working at the AI for Good Research Lab, geospatial scientists Anthony Ortiz and Caleb Robinson created a comprehensive map showcasing the distribution of wind farms and solar farms worldwide. Crucially, this map also incorporates landcover designations, allowing for the identification of potential conflicts between land use for renewable energy generation and agricultural activities. This resource remains updated thanks to Planet's frequent mapping of the globe.

Training and Education In regions with diverse linguistic backgrounds, communication between farmers and agricultural experts can be challenging. However, AI-powered natural language processing (NLP) systems can bridge language barriers by enabling farmers to communicate and receive information in their native languages. This can help farmers access critical information about weather forecasts, market prices, and best practices for crop management.

In Kenya, for example, the Alliance for a Green Revolution in Africa (AGRA) has vast databases of agricultural information that can greatly benefit local farmers. However, the challenge lies in bridging the language barrier between the farmers and the databases. Currently, an intermediary, often a person in a village or a group of villages, acts as the go-between, translating questions from farmers into a language the databases can understand and relaying the answers back to the farmers.

But since mobile phones are widely used in Kenya, even in rural areas, the ability to reach farmers directly and have natural conversations is tempting. One organization leading the charge in Africa is Masakhane,[26] a community organization of African-created AI researchers passionate about advancing NLP and machine learning in Africa. They are committed to promoting inclusivity, diversity, and open science while addressing real-world problems, such as language preservation, healthcare, education, and agriculture. With the widespread use of mobile phones, even in rural areas of Kenya, the potential to directly communicate with farmers is immense, thanks to AI.

The use of AI-powered NLP systems in agriculture has the potential to transform the way farmers communicate, access information, and manage their crops. It is a game-changer for regions with diverse linguistic backgrounds, making critical agricultural information more accessible and empowering farmers to make informed decisions for their livelihoods. The power of AI is not just about technological advancements but also about addressing real-world challenges and promoting inclusiveness in underserved communities.

Medicine

Medical Imaging

For thousands of years, doctors relied solely on their senses to diagnose ailments. Assisted by some rudimentary instruments, they embarked on a guessing game, where "poke and prod" was the norm, and patients often flinched in pain. Hippocrates, the "Father of Medicine," was among the first to emphasize the importance of observing and examining patients to determine their ailments.

Doctors would palpate the patient's body, searching for telltale signs of disease. Swollen glands, tender spots, and abnormal growths were all discovered through this tactile exploration. They would scrutinize a patient's skin color, the whites of their eyes, and the state of their tongue for any signs of illness. In some cases, physicians even tasted their patient's urine, as the presence of sugar could indicate diabetes. This practice, while undoubtedly unpalatable, was a testament to the lengths doctors would go to diagnose their patients accurately.[27]

The sense of smell played a surprisingly significant role in ancient diagnostics. A physician would often take a deep whiff of a patient's breath, sweat, or bodily excretions to detect any unusual odors that could signal illness. In fact, the sweet, fruity smell of a diabetic's breath or the ammonia-like scent of a patient with kidney failure could provide valuable clues to their condition.

Wilhelm Conrad Roentgen's discovery of X-rays just before the 20th century gave doctors a new superpower. They suddenly had a way of peering inside the human body without making a single incision. They were admitted to a whole new realm of visual observations.

This superpower came with a cost, though. X-rays are ionizing radiation, meaning that they knock electrons off atoms. Imagine this happening to your body; suddenly, it doesn't seem so non-invasive. Observing the effect of X-rays on the human body,[28,29] scientists began looking for other ways to see inside the body. After decades of improving X-ray technology, medical imaging technology rose sharply. In the 1950s, researchers took a technique developed during World War II to discover submarines and decided it would be perfect for female reproductive organs during pregnancy. This was ultrasound, which was much safer than X-rays. Soon after, magnetic resonance imaging (MRI) was developed. These machines create a magnetic field—too weak to cause any harm—that aligns the nuclei of hydrogen atoms in the body. Then these nuclei are bombarded by radio waves, which causes them to emit their own weak radio signals. These signals are picked up by a machine that creates an image of the body.

After that came positron emission tomography (PET) and single-photon emission computed tomography (SPECT) scans. In these methods, a small amount of radioactive tracer is injected into the patient. It is designed to accumulate in the organ of interest, and there emit gamma rays. Gamma-ray detectors arranged around the patient show the real-time working of the organs. These sound even more dangerous than X-rays, but the actual amount of radiation is much less than that produced by standard X-rays. And they are immensely useful in detecting cancers and brain anomalies.

Figure 3.3 shows what some of these images look like. Radiologists have become attuned to the subtle variations in textures, contours, and densities within the images, and can identify tumors, decreased blood flow, or a number of other irregularities. Their pattern recognition abilities have been honed through years of extensive training akin to chess grandmasters recognizing their strengths and weaknesses upon a glance at the chess board.

Given AI's incredible aptitude for image analysis, radiologists must wonder if their profession will go the way of the horse-and-buggy. Amidst their trepidation, they should find solace in that they will not be displaced entirely, but it is reasonable to anticipate their daily routines may change.

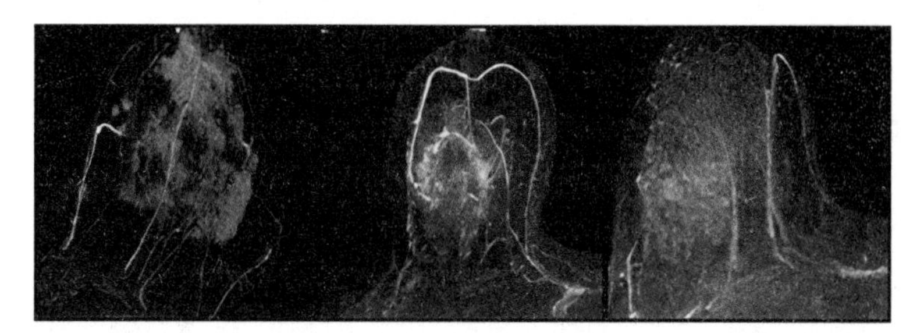

Figure 3.3 MRI images of breast cancer

The fear of job displacement accompanies the rise of every new technology. When MRIs were introduced, similar concerns were raised. The level of detail and precision offered by MRIs compared to traditional CT scans was radical, leading to speculation that radiologists might become obsolete. Surprisingly, the opposite occurred—the demand for radiologists increased.

AI algorithms can scrutinize medical images with unparalleled precision and speed. Learning from a massive amount of medical image data, AI models can become experts at detecting patterns and identifying anomalies. This means that doctors can get more accurate diagnoses of diseases, possibly at earlier stages when treatment is most effective. In this case, AI acts like a decision assistant, an extra set of very smart eyes that helps radiologists with their diagnoses.

AI algorithms can also automate and streamline many aspects of medical imaging workflows. Image acquisition, segmentation, and analysis are what they would excel at. For example, AI-powered image reconstruction techniques can enable faster and higher-quality image acquisition, reducing radiation exposure for patients in the case of X-rays and CT scans. AI algorithms can also assist in automating routine tasks, such as image annotation and report generation, freeing up radiologists' time to focus on more complex and critical cases.

Currently, this is still a dream. However, it is a realistic, attainable dream and one that can come true if radiologists work to make it happen. The AI for Good Lab has worked on several projects developing AI tools to detect pancreatic cancer. Dr. Eliot Fishman, a radiologist at Johns Hopkins University, has spoken extensively about AI and radiology. He insists that in the end, it is the patients who come out ahead.[30] Pancreatic cancer, known for its dismal survival rates, is Dr. Fishman's specialty. He has referred to the pursuit of improved pancreatic cancer detection and treatment as "our Manhattan Project," a bold endeavor to overcome the challenges associated with this deadly disease.

But while there has been progress on some diseases, we do not have AI algorithms that can detect cancer in other organs at the same level of accuracy. This will require similar efforts to the ones undertaken by Dr. Fishman and his colleagues, including help from radiologists.

Pathology

The other field in medicine that seems ripe for transformation is pathology, the science of determining what ails a person. A pathologist is what you get when a detective goes to medical school. They put slides under microscopes, do lab tests on blood or urine, and look for patterns and features on cells, for metabolic markers, and genetic mutations to figure out diseases.

From an AI perspective, the work can be abstracted into tasks such as pattern matching and visual anomaly identification. Microscope slides can be digitized into high-resolution images. Blood parameters such as glucose levels and liver and kidney markers can be transformed into data that machine learning algorithms can digest. Genetic data acquired through DNA sequencing can be converted into sequences of base pairs. Then, AI can be used to look for signs of various diseases.

Traditionally considered a behind-the-scenes specialty, pathology now finds itself at the confluence of medicine and artificial intelligence. It is a rich field for AI startups that can develop models and devices to assist physicians making diagnoses. The US Food and Drug Administration (FDA) has approved AI tools for prostate screening, cervical smear screening, and cancer detection,[31] and many are lining up to get the coveted and necessary CE (Conformité Européenne) mark from European regulators.[32,33]

With AI making inroads into medicine through radiology and pathology, how will it be regulated? Automation has been part of medical technology before, but now we stand on the verge of a different kind of innovation. The rapid pace of medical knowledge advancement already requires radiologists and pathologists to continually update their knowledge through continuing education courses and certification renewals. Medical devices also need to be monitored and updated regularly. One can view AI as just another tool with the physician remaining the final decision-maker under existing regulations. But if the AI tool itself is flawed, say with biased training data, is it the physician or the AI model developers who are liable for its mistakes?

Then, for AI systems to make a significant impact, they will need access to large amounts of data. That is usually possible if data from several institutions can be combined. Setting aside the practical problems of combining data gathered with different protocols and devices, questions of data privacy, confidentiality, and other regulatory issues present significant barriers to implementation.

We will continue exploring the questions of AI regulation in Chapter 6.

Education

Education is the fourth of the UN's 17 sustainable development goals, recognized as a universal human right. It is essential for the growth of civilized societies, the well-being of our species, and the health of our shared global home. Yet, a stark contrast exists between this ideal and the reality in regions like sub-Saharan Africa. If you're born there, there's a one in five chance you won't receive a basic education. The sheer number of children out of school in this region even outnumbers the rest of the world combined. Without proficiency in reading, writing, and mathematics, these children are destined to be stuck in their harsh economic circumstances.

The promise of AI in making education more equitable and reliably nurturing each person's potential is still aspirational. Efforts to create these systems are under way, but the hurdles they face are formidable. Schools often lack the necessary infrastructure. Teacher training and support systems will require a complete overhaul. And then there are the pressing concerns about the students' privacy. The journey to harness AI's promise in education is fraught with challenges, but the potential rewards are too great to ignore.

More so than AI in medicine, environmental challenges, and other areas we've discussed here, AI in education is the least developed. Attempts have been made in several applications, but they all fall short in some way or other. Let's review some of these applications.

An example of using AI in the form of a virtual mentor is Jill Watson, an AI teaching assistant created and used at Georgia Tech. Jill is a cousin of IBM's Watson and acts as a chatbot, answering students' questions about the course, syllabus, logistics, and so on, just as a teaching assistant would. In primary and secondary schools, where teachers are highly overworked, such a system might offer them some relief by taking on some of their more mundane tasks. Grading multiple-choice tests is easy, but AI could potentially grade essays. And as more education moves to the laptops and online, AI can analyze student activity data, such as time spent on tasks, completion rates, and performance, to identify trends and patterns. This information can help educators identify at-risk students and refine their teaching strategies.

One class of tasks that still seems far away is the takeover of actual tutoring. Proponents of intelligent tutoring systems would like to see software that can adapt to a student's learning style. It would modify itself in real time based on interaction with the student providing personalized feedback and guidance. The AI would not make up the content or curriculum itself, which would be curated by human teachers. Such systems do exist, like SquirrelAI,[34] developed in Shanghai, China and reportedly used in learning centers in several cities in China.[35]

Final Thoughts

We stand at a unique juncture in the story of humanity. Over the past century, we've faced challenges and threats unparalleled in their magnitude, from the threat of nuclear holocaust to global environmental degradation. At the same time, we have created technologies that may equip us with the means to address these challenges. The AI examples discussed in this chapter are but a beginning.

As successful applications of AI grow and proliferate, they will become an essential ingredient in our efforts to make a better world for future generations.

Artificial intelligence by itself is not a panacea for all our problems. It is a tool—one whose effect will hinge on the collective values and ethos of our global society. The way forward for human society is likely in our minds, attitudes, and behaviors. These will shape how we apply AI for good.

References

1. https://observer.com/author/michael-sainato. Stephen Hawking, Elon Musk, and Bill Gates Warn about artificial intelligence. *Observer* (2015). at <https://observer.com/2015/08/stephen-hawking-elon-musk-and-bill-gates-warn-about-artificial-intelligence/>
2. Google CED: AI impact to be more profound than discovery of fire, electricity - CBS News. at <https://www.cbsnews.com/video/google-ceo-ai-impact-to-be-more-profound-than-discovery-of-fire-electricity/>
3. Henson, B. Cruel echoes of a 2010 disaster in Pakistan's catastrophic 2022 floods » Yale Climate Connections. *Yale Climate Connections* (2022). at <http://yaleclimateconnections.org/2022/08/cruel-echoes-of-a-2010-disaster-in-pakistans-catastrophic-2022-floods/>
4. Big Tech and Artificial Intelligence Can Support Early Warnings for All. (2023). at <https://public.wmo.int/en/media/news/big-tech-and-artificial-intelligence-can-support-early-warnings-all>
5. Nasir, M., Sederholm, T., Sharma, A., Mallu, S. R., Ghatage, S. R., Dodhia, R. & Ferres, J. L. Dwelling type classification for disaster risk assessment using satellite imagery. *arXiv preprint arXiv:2211.11636* (2022).

6. IFRC. Turkiye - Earthquake (MDRTR004) | Operational Update #2. (2023). at <https://www.ifrc.org/media/52045>
7. Fair, C. C., Kuhn, P. M., Malhotra, N. & Shapiro, J. N. Natural disasters and political engagement: Evidence from the 2010–11 Pakistani floods. *QJPS* **12,** 99–141 (2017).
8. Tatem, A. J. WorldPop, open data for spatial demography. *Scientific Data* **4,** 1–4 (2017).
9. "Tantalizing clue" marks end of Amelia Earhart expedition. *Culture* (2019). at <https://www.nationalgeographic.com/culture/article/tantalizing-clue-marks-end-amelia-earhart-expedition>
10. How to save rainforests with satellites? Highlights of the NICFI Satellite Data Program. *NICFI* (2022). at <https://www.nicfi.no/current/how-to-save-rainforests-with-satellites-highlights-of-the-nicfi-satellite-imagery-program/>
11. Ortiz, A., Tian, W., Sherpa, T. C., Shrestha, F., Matin, M., Dodhia, R., Ferres, J. M. L. & Sankaran, K. Mapping glacial lakes using historically guided segmentation. *IEEE Journal of Selected Topics in Applied Earth Observations and Remote Sensing* **15,** 9226–9240 (2022).
12. Gutro, R. NASA Covers Wildfires Using Many Sources. *NASA* (2015). at <http://www.nasa.gov/mission_pages/fires/main/missions/index.html>
13. Tylor, G. Top 11 Innovative Startups Applying AI for Wildfires Detection in 2022 - exci. (2022). at <https://www.exci.ai/top-11-innovative-startups-applying-ai-for-wildfires-detection-in-2022/>
14. Bi, K., Xie, L., Zhang, H., Chen, X., Gu, X. & Tian, Q. Accurate medium-range global weather forecasting with 3D neural networks. *Nature* 1–6 (2023). doi:10.1038/s41586-023-06185-3
15. Ebert-Uphoff, I. & Hilburn, K. The outlook for AI weather prediction. *Nature* (2023). doi:10.1038/d41586-023-02084-9

16. Jiao, P. & Alavi, A. H. Artificial intelligence in seismology: Advent, performance and future trends. *Geoscience Frontiers* **11,** 739–744 (2020).
17. Yabe, T., Jones, N. K. W., Rao, P. S. C., Gonzalez, M. C. & Ukkusuri, S. V. Mobile phone location data for disasters: A review from natural hazards and epidemics. *Computers, Environment and Urban Systems* **94,** 101777 (2022).
18. McLaughlin, T. The plan to save India from disinformation. *The Atlantic* (2018). at <https://www.theatlantic.com/international/archive/2018/09/fighting-whatsapp-disinformation-india-kerala-floods/569332/>
19. "We live in fear": Facing dry times, rural South Africans rethink water. *Reuters* (2019). at <https://www.reuters.com/article/us-water-safrica-scarcity-idUSKCN1T70KH>
20. Becker-Reshef, I., Vermote, E., Lindeman, M. & Justice, C. A generalized regression-based model for forecasting winter wheat yields in Kansas and Ukraine using MODIS data. *Remote Sensing of Environment* **114,** 1312–1323 (2010).
21. Krishnan, V. G., Deepa, J., Rao, P. V., Divya, V. & Kaviarasan, S. An automated segmentation and classification model for banana leaf disease detection. *Journal of Applied Biology and Biotechnology* **10,** 213–220 (2022).
22. African Development Bank. Introduction to Mini-Grids. *Energy for Impact* at <https://greenminigrid.afdb.org/how-it-works/help-desk-developers-and-operators/introduction-mini-grids>
23. World Bank. *Mini Grids for Half a Billion People: Market Outlook and Handbook.* (World Bank Publications, 2021).
24. Solar Mini Grids Could Sustainably Power 380 million People in Africa by 2030 – If Action Is Taken Now. *World Bank* at <https://www.worldbank.org/en/news/press-release/2023/02/26/solar-mini-grids-could-sustainably-power-380-million-people-in-afe-africa-by-2030-if-action-is-taken-now>

25. Baruch-Mordo, S., Kiesecker, J. M., Kennedy, C. M., Oakleaf, J. R. & Opperman, J. J. From Paris to practice: Sustainable implementation of renewable energy goals. *Environmental Research Letters* **14,** 024013 (2019).
26. Orife, I., Kreutzer, J., Sibanda, B., Whitenack, D., Siminyu, K., Martinus, L., Ali, J. T., Abbott, J., Marivate, V. & Kabongo, S. Masakhane—Machine Translation for Africa. *arXiv preprint arXiv:2003.11529* (2020).
27. Rutkow, I. M. *Empire of the Scalpel: The History of Surgery.* (Scribner, 2022).
28. Moore, K. *The Radium Girls: The Dark Story of America's Shining Women.* (Sourcebooks, Inc, 2017).
29. Nelson, C. *The Age of Radiance: The Epic Rise and Dramatic Fall of the Atomic Era.* (Scribner, 2014).
30. Will artificial intelligence replace doctors? AAMC at <https://www.aamc.org/news/will-artificial-intelligence-replace-doctors>
31. US Food and Drug Administration. Artificial intelligence and machine learning (AI/ML)–enabled medical devices. *FDA* (2022). at <https://www.fda.gov/medical-devices/software-medical-device-samd/artificial-intelligence-and-machine-learning-aiml-enabled-medical-devices>
32. Hale, C. Pathology AI developer Paige scores CE marks in breast, prostate cancers. *Fierce Biotech* (2020). at <https://www.fiercebiotech.com/medtech/pathology-ai-developer-paige-scores-ce-marks-breast-prostate-cancer>
33. Park, A. Aiforia lines up another AI pathology approval in Europe, now for diagnosing breast cancer. *Fierce Biotech* (2022). at <https://www.fiercebiotech.com/medtech/aiforia-lines-another-eu-approval-ai-pathology-model-diagnose-breast-cancer>

34. Hao, K. Born in China, taught by AI. *MIT Technology Review* **123,** 24–29 (2020).

35. Online Education. Jill Watson: Using AI and Adaptive Learning to Help Teachers and Students. at <https://www.onlineeducation.com/features/ai-teaching-assistant-jill-watson>

4

AI for Good: Pursuit of Scientific Knowledge

"Science knows no country, because knowledge belongs to humanity, and is the torch which illuminates the world."

– Louis Pasteur

"Nature is the source of all true knowledge. She has her own logic, her own laws, she has no effect without cause nor invention without necessity."

– Leonardo da Vinci

CHAPTER 3 COVERED how AI can affect issues at a societal level. Here, we continue that exploration, focusing on AI's effect on our environment and our pursuit of scientific knowledge.

Scientists have been among the first to jump at the opportunities that AI offers to supercharge their research. AI flourishes particularly in fields where new correlations and insights are waiting to be found hidden in voluminous data, such as genetic and protein research, astronomy, pharmacology, and materials science. While many of these fields have built specialized artificial intelligence tools, the promise of foundational models, discussed in Chapter 2, will expand the reach of AI to many other fields.

In this chapter, we will look at how scientists incorporate AI into their work. We will look at how research scientists in biology and conservation have made AI part of their arsenal of research tools with which to study endangered populations. Then, we explore how AI has revolutionized the study of proteins, and we end with how AI helps us find new planets and possibly other types of intelligence.

Biodiversity

When we consider the mind-boggling array of life on Earth, it might be hard to imagine how one species can have such an outsize influence on a planet to change its very nature. We cunning, resourceful humans have managed to trigger an extinction of living species that is comparable to the biggest die-offs that have occurred in Earth's long history.[1]

However, one might argue, isn't that just part of evolution? The world is in a constant state of change, and for us to freeze conditions at one geological moment in time seems naïve and impractical. The state of life on Earth at the time of the dinosaurs was very different. Since then, the great beasts that once walked the Earth and the leviathans that swam the seas are no longer around. The Anthropocene, or age of human influence, represents just one link in the extensive chain of Earth's historical eras. Why should the current state of affairs concern us so much?

While appealing in its simplicity and human-centric view, such an argument is fatalistic and unrealistic. What is concerning about this era is the breakneck speed at which we are hurtling toward change, leaving little time for adaptation, whether mental, spiritual, or even physical. We may find ourselves in a world that may be worse for our survival, let alone for countless other species. As historian Yuval Noah Harari cogently pointed out, we have been the cause of extinction everywhere we have gone, from the Americas to Madagascar to Australia.[2] Even if the moral and ethical reasons for changing our impact on the world are insufficient to change our behavior, surely a biological reason, our survival, may motivate a change in our behavior.

Beluga Whales

An easy way to understand how fraught the changes in an environment can be is to see what happens when a keystone species is endangered or even removed.

Keystone species are major players in an ecosystem, the ones that have a disproportionate impact on keeping everything in balance. They are like the linchpin in a machine or, as the name implies, the keystone in an arch. If you take them out, the whole thing falls apart.[3]

Before we get to beluga whales, let us consider beavers to illustrate the point. Beavers live near moving water and are considered a keystone species because of the way they modify their environment. They build dams that serve as habitats and enable them to obtain their food. These dams create ponds and wetlands that support a variety of other species, from fish and amphibians to birds and insects. Without beavers, these habitats would be drastically altered, and many species that rely on them would suffer.

Now, take your typical domestic cat. It may seem to rule your house and is likely to be one of the leading causes of bird and small mammal mortality in your neighborhood, which can have ripple effects on the food web and ecosystem dynamics. That does not necessarily make it a keystone species. Wild cats, however, such as lions, tigers, and leopards can be considered keystone species. These large predators have a major impact on the populations of their prey, and their absence can lead to overgrazing or other imbalances in the ecosystem. For example, in African savanna ecosystems, lions are considered keystone species because they help control the populations of herbivores, such as zebras and wildebeest, which in turn helps regulate the growth of vegetation and maintain the balance of the ecosystem.[4]

Like a canary in a coal mine, a keystone species can reveal the state of its environment. A thriving keystone bodes well for the ecosystem. But when a keystone species is struggling, it is a red flag for the environment; it is a signal from nature that something is amiss. So, monitoring the health of keystone species can provide early indicators of the well-being of our planet.

Conservationists monitor animal populations using several different techniques. Individuals could be radiotagged, or microphones or cameras could be hidden in the wild, on trees or underwater, and their sounds captured. Or researchers might just sit and make themselves comfortable in a spot and count how many individuals they see in a day. Cameras and microphones generate vast amounts of data, millions of photos or videos, or hours of sound recordings. Traditionally, a good portion of college students' lives are spent poring over these data, looking for signals. However, future students will likely use artificial intelligence to do this job for them. In fact, scientists at the National Oceanic and Atmospheric Administration (NOAA) are already using this technology for beluga whale conservation.

Beluga whales in Alaska are facing a serious problem. Their populations have become greatly reduced to the point where they are considered "depleted." This isn't just a regular word—it has a specific meaning in the legal language of the Marine Mammal Protection Act. It signifies that the beluga whale population is no longer able to sustain itself in a healthy way. In the late 1970s, these whales were already in a vulnerable state, classified as endangered, and their numbers were around 1,300. However, as time has passed, their population has drastically decreased. As of 2023,[5] there are less than 300 beluga whales remaining in these Alaskan waters.

This is extremely concerning for the sea ecosystems of Alaska and the Arctic because beluga whales are considered a keystone species. Their decline has been linked to changes in the abundance of fish, shrimp, and other bottom-dwelling creatures, their prey. Known as trophic cascade, this will lead to greater vulnerabilities to environmental disturbances.

In 2017, a NOAA team led by Manuel Castellote began monitoring the Cook Inlet population. They submerged hydrophones and picked up recordings of beluga whale sounds. Then, his team spent countless hours listening to the recordings of sounds made by the whales. In this way, they are able to estimate how many whales there are in Cook Inlet.

This census conducted through bioacoustics is non-invasive and passive. The subsequent analysis, after the recordings, is time-consuming and repetitive. It is perfect for artificial intelligence. Within a matter of weeks, AI experts from the AI for Good Research Lab built a very reliable AI system that determined the exact time points in the recordings where a beluga whale call could be found.[6,7] It was good at rejecting calls that were not beluga whale calls and could capture almost all calls.

This was an obvious merger of science and AI. The AI models that exist can be readily adapted to other species.

In fact, with current advances in foundational models (see Chapter 2), scientists will soon be able to use interfaces such as chatbots to analyze their bioacoustic data.

Rainforests

The Amazon, Congo, and other rainforests have been variously described as the lungs or the beating heart of the Earth. These metaphors are inspired by their function; they literally breathe life into our planet by pumping out enormous amounts of oxygen. They may also play a significant role in regulating climate, particularly by influencing atmospheric circulation and rainfall patterns.[8] More than a source of oxygen, rainforests are likely more important in the role they play in regulating carbon dioxide, a greenhouse gas, and as treasure troves of biological diversity.

When the Amazon was burning in 2019, the danger of losing one of the world's most important and diverse biological treasures captured the attention of a significant portion of the world's population, including heads of state and celebrities. Most wildfires in these wet regions are often set intentionally[9] by groups interested in clearing the land for mining or agriculture. The enormity of the fires in 2019 reignited the fear that the Amazon could degrade into steppe-like drylands, with immeasurable harm to the whole world.[10]

Meanwhile, in the Democratic Republic of the Congo, despite the end of the civil war, militia groups were slaughtering mountain gorillas and other wildlife. This was the consequence of another economic conflict between those who wanted to preserve the UNESCO World Heritage Site that is Virunga National Park and the people who wanted to exploit the mineral and other natural resources for economic livelihoods.

The vastness of the Amazon and Congo makes them incredibly difficult to monitor and protect. The Amazon, for example, extends across nine countries and territories: Brazil, Bolivia, Peru, Ecuador, Colombia, Venezuela, Guyana, Suriname, and French Guiana. The Congo Basin, in comparison, stretches across six countries: Cameroon, Central African Republic, Democratic Republic of the Congo, Republic of the Congo, Equatorial Guinea, and Gabon. In the Congo, mountain gorillas and other wildlife are decimated by militia groups who want to discourage tourism.[11]

Monitoring the Amazon is beyond the combined capabilities of the nine countries across which it sprawls. The only highways through these vast regions are those built by logging and mining companies, which encourage more deforestation. Recognizing the need for a coordinated response, Colombia led the region in an agreement called the Leticia Pact.[12] In 2023, the members of the Pact saw a demonstration of how AI could help further their efforts through satellite imagery, bioacoustics, and camera traps.

In the Congo Basin, a similar coming together of governments seems unlikely. However, the same AI technologies, in the hands of individual governments, may still lead to the successes the Amazon is hopefully headed for.

Satellites

Satellites orbiting the Earth at altitudes of hundreds of miles are able to capture images of the surface at remarkable resolution. At 10m per pixel, images captured in the visible spectrum can help identify fires. At that resolution, deforestation can also be identified. At higher resolutions, say 3m per pixel, analysts can potentially detect more, such as illegal mining or illegal settlements.

But satellite images cannot show us what lies beneath the tree canopy. So, instead of regular, visual images, we go down the frequencies on the electromagnetic spectrum and use lasers and radar. The laser technology is known as LiDAR, which stands for Light Detection and Ranging, and the radar technology is known as Synthetic Aperture Radar (SAR).

LiDAR is a technology that uses laser pulses to measure the distance between the satellite and the Earth's surface. These pulses are intense but of very short duration, in the order of nanoseconds, which is billionths of a second. If these laser pulses hit you or animals in the forests as they are doing their scans, they would do no harm since the intensity, usually in the near-infrared part of the spectrum, just isn't strong enough. Autonomous vehicles use LiDAR to navigate.

What these lasers can do is generate detailed 3D maps of forest structure, including the heights of trees and other vegetation, density of the forest canopy, and the biomass of individual trees. They are also used to estimate the amount of carbon stored in trees. The LiDAR data is usually combined with ground-based measurements.

Scientists and researchers walk around forests and fields doing exhaustive measurements that can be used to calibrate the satellite LiDAR data. Then, mathematical models are employed to interpret the data from the satellites and estimate carbon stocks. These models take into account factors such as tree species, age, and environmental conditions to calculate the carbon content of forests based on satellite measurements.

Complementing the visible spectrum and LiDAR is radar. Like LiDAR, radar is not confounded by cloudy days or nights. This is particularly useful for monitoring rainforests, which, unsurprisingly, are often covered by clouds. Just as lasers can help create high-resolution maps beneath the canopy, so can radar.

In addition to monitoring the health of the rainforests, AI can also play a role in combating illegal activities that contribute to deforestation. Machine learning algorithms can be used to analyze data from a variety of sources, including satellite imagery, LIDAR data, and bioacoustic recordings, to identify potential areas of illegal logging, mining, or poaching. By pinpointing these areas, authorities can focus their enforcement efforts and work to protect the fragile ecosystems of the rainforests.

AI technologies can also be used to support reforestation efforts. For example, drones equipped with AI-powered cameras can identify suitable areas for planting new trees and even help monitor the growth and health of the newly planted saplings. This can significantly improve the efficiency and effectiveness of reforestation efforts, helping to restore these vital ecosystems more quickly.

Bioacoustics

Like Manuel de Castellote's whale sounds in Alaskan waters, forest sounds can also be used to protect and restore ecosystems. In 2021, a number of researchers from the AI for Good Research Lab teamed up with Sieve Analytics to identify bird, frog, and insect species in the Yunque National Forest in Puerto Rico. The researchers at Sieve Analytics had mounted microphones at 700 sites in the forest, and they had been programmed to record 1-minute audio every 10 minutes 24 hours per day for 1–2 weeks per sampling site.

The first pass at AI models to identify animal species revealed some difficulties. The forest tends to be noisy, and birds don't take turns when communicating. Some of the less numerous species hardly feature in the recordings even with microphones positioned at hundreds of sites.

My team faced similar challenges when listening to sounds gathered at the Orinoquía Forest in Columbia. Researchers at Humboldt University tagged the sounds of birds they recognized with their scientific names. AI scientists at the AI for Good Research Lab transformed the audio into spectrograms, which are graphs of call frequencies. This transformation made the problem more tractable by converting it into a question of image classification.

AI's impact on biodiversity has enabled conservationists to collect and analyze previously unfeasible data. With inexpensive technology, microphones and cameras can be strung out across large areas. The enormous amount of data that is subsequently collected can be analyzed intelligently in its entirety.

Camera Traps

A tool that has ascended to prominence in conservation circles is the digital camera. When they are weatherproofed, have a portable source of power such as built-in batteries, and a way to transmit or store images and video, they give rangers eyes in specific locations they cannot patrol every day.

They are usually triggered by movement, which results in some incredible and beautiful captures of wildlife, as shown in Figures 4.1 and 4.2. You might have these cameras in your home, pointing at your backyard or the front door, alerting you when a person appears. Some domestic systems will even tell you when they detect a pet or some other creature or when packages are delivered.

Figure 4.1 Camera trap image of an ocelot

Source: Missouri camera traps dataset hosted on LILA.[13]

Figure 4.2 Camera trap image of a wild turkey

Source: ENA24 dataset hosted on LILA.[14]

These "traps" can capture thousands of photos each when used in the wild. The majority of them, sometimes 99%, are empty—they do not have animals in them. As in bioacoustics, AI can assume the role of an undergraduate or graduate research assistant and can sift through these images very quickly, identifying those with an animal that a conservationist can then examine. This application of AI can save researchers an enormous amount of time, freeing them to focus on more complex aspects of their research.[15] The models become such experts at detecting wildlife that they'll even find the birds camouflaged in Figures 4.3 and 4.4.

One of the most effective and widely used AI tools for camera traps was created by a former Microsoft researcher, Dan Morris. With his collaborators, he developed an

Figure 4.3 Camera trap images of hard-to-detect birds in the Amazon rainforest

open-source tool known as Megadetector.[16,17] A job that seems easy for humans—is there an animal in this photo—requires substantial training for an AI. But after the AI has learnt some representations of animals, it can work at the speed of the CPU on your laptop to rapidly sift through large collections of these photos.

Figure 4.4 Camera trap images of hard-to-detect birds, with bounding boxes superimposed by an AI model. On the top is genus Ortalis (chacalacas). On the bottom is Genus Columbina (doves)

The potential applications of camera traps are manifold. They could serve as an effective mechanism to detect and deter poaching activities. For example, in Borneo's dense jungles, the Borneo Orangutan Survival Foundation has camera traps monitoring orangutans' activities. In South Africa, an all-female ranger team called the Black Mambas Anti-Poaching Unit[18,19] uses camera traps within the Balule Nature Reserve to protect rhinos and other wildlife from poachers. In India, the Wildlife Conservation Society (WCS) has partnered with the Indian government to deploy camera traps in protected areas such as the Nagarahole and Bandipur National Parks.[20,21] Their mission is to safeguard tiger populations in these areas, and they use their cameras to monitor wildlife movement. Similarly, in Tanzania, the WCS is working with the country's national parks organization to keep a protective eye on lion populations.[22] These camera traps are extremely useful for conservationists, even if there is a chance of vandalism by poachers, because they provide eyes where rangers cannot always patrol.

Proteomics

When you crack an egg into a frying pan, the semi-colorless, viscous liquid around the yolk quickly transforms into a solid, white mass. And when you leave milk out, you'll find it has become clumpy and smelly. It has curdled, meaning solids have separated from the liquid. Both of these transformations happen because proteins are breaking down and then reassembling in new ways. The egg proteins break down because heat causes them to unfold, or denature. This denaturation exposes new surfaces in the shape of the molecule, allowing proteins to bond with each other more tightly and forming a solid network that we see as a cooked egg white.

A similar metamorphosis happens to the forgotten milk left on your kitchen counter. In this case, it is not the heat that denatures milk proteins. The responsible agent here is lactic acid, which is produced by bacteria that naturally occur in milk. Lactic acid causes the proteins in milk to unfold and aggregate, leading to the curdling effect.

When Cleopatra bathed in sour milk, or the modern beautician gives you a chemical peel, proteins on the surface of the skin—collagen being one of the most common ones—are broken down. Scientists in the 18th and early 19th centuries suspected "animal substances" made up skin, blood, and muscles. They found a similar substance in wheat flour, which we now call gluten.

When they could decipher the molecular composition of these animal substances, they were surprised to find that all these different varieties of "animal substances" were strikingly similar, consisting of carbon, hydrogen, nitrogen, and oxygen. They speculated that one primary substance occurred in plants and traveled to animals by eating these plants. This primary substance was named protein, from the Greek "proteios," which roughly means "of the first quality."

But it took over a hundred years for scientists to discover the structure of proteins. In an oft-repeated occurrence in the history of science, it took the money of a major grant-making organization to turn a brilliant mind to this problem. The Rockefeller Foundation was the early 20th century's version of today's Gates Foundation, the largest philanthropic organization of its time. Its grants attracted Caltech chemist Linus Pauling to a new field of research that used new imaging techniques such as

X-ray crystallography to investigate the shape of biological molecules.

The results of any scientific discovery are rarely due to just one person. Building on the ideas of a thriving research community, Pauling's research came to a crescendo in 1951. In a series of seven papers published in the Proceedings of the National Academy of Sciences (PNAS),[23,24] Pauling and his co-authors, Robert Corey and Herbert Branson, provided the first detailed descriptions of the structures of proteins. The big prize, structure of the DNA molecule, eluded him, however. The eventual discoverers of that molecule's structure, Francis Crick and James Watson, used Pauling's ideas to win the prize.

Scientists now understood that proteins, these animal substances that animated skin, blood, and muscles, were made up of amino acids created in the cytoplasm of cells. The proteins' fantastic structures are determined by DNA. This DNA provides instructions to messenger molecules known as RNA, which then guide the assembly of the 20 amino acids in the cytoplasm. Unlike the iconic, double-helix shape of the DNA molecule, protein molecules can take on shapes that are corkscrews, globular, long strands, all folded in ways that enable formations of skin, muscles, etc. Figure 4.5 illustrates the complex shape of proteins.

Just as a genome indicates the complete set of genes that a person possesses, a proteome describes the complete set of proteins in an organism. A person has between 20,000 to 25,000 genes, and between 10,000 and billions of proteins.

This is surprising and a reason why proteomics is a major field of research. Millions of new proteins are

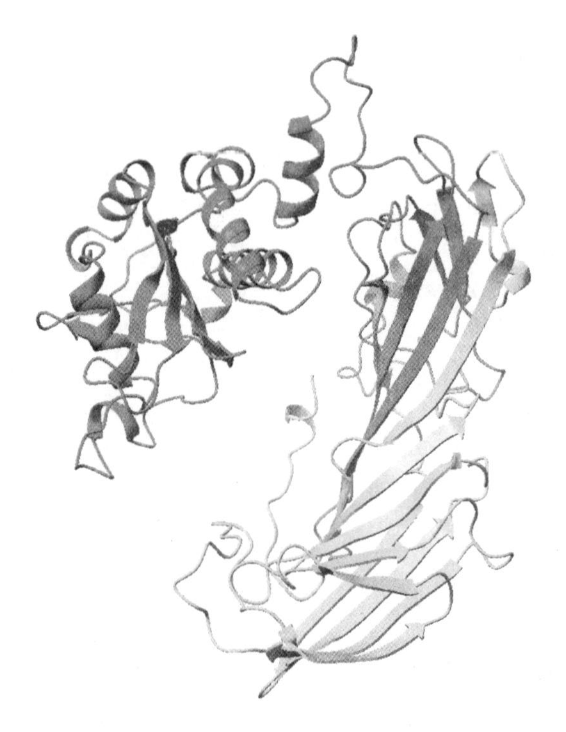

Figure 4.5 An example of the complex structure of proteins. The 3D structure of the protein AP5M1, identified using Alphafold. Three domains are identified by color, showing parts of the protein that can fold independently of the others.

Credit: Wikimedia Commons (user: Tobithias)

discovered every year, but without an understanding of their function, The way proteins contort a single string of molecules into a complex 3D shape, known as the protein folding problem, became a scientific research endeavor on par with the most famous of scientific quests such as Fermat's Last Theorem and the discovery of DNA.

Then, like a bull crashing through barriers, an AI algorithm developed by DeepMind suddenly made discovering the 3D structure of proteins possible. Like image recognition models developed in response to a competition, AlphaFold was released at a competition called the Critical Assessment of Protein Structure Prediction (CASP).[25]

AlphaFold has made possible research that was believed to be so difficult that most scientists didn't even consider it. Development continues at a relentless pace. Within a couple of years, AlphaFold 2 was released, and at about the same time, the Baker Lab at the University of Washington in Seattle released RoseTTAFold. RoseTTAFold utilizes similar deep learning techniques to AlphaFold 2 but can also predict structures of complex protein assemblies.

The incredible impact of these models cannot be overstated. Their development has triggered a revolution in proteomics, which was hindered by the complexity of the protein folding problem. Now, the speed at which new protein structures can be predicted and potentially understood has increased exponentially, accelerating scientific understanding at an unprecedented rate. What we'll see from this advance is a much greater understanding of diseases like Parkinson's, Alzheimer's, and several others that are associated with protein folding. We will see drugs designed to fit targeted proteins.

Astronomy

Astronomers quickly realized that AI's ability to spot objects in complex photos and to find patterns in complex data was remarkably well suited to their tasks. Contrary to popular notions of a lonely astronomer sitting at the top of a hill in a domed observatory peering through a telescope, astronomers tend to spend much of their time in front of a computer, analyzing data captured digitally by their telescopes. They might still be lonely.

AI works well on huge data, particularly at tasks that need to be scaled up. Astronomy provides data in droves. When a cell phone takes a picture, its size might be on the order of several megabytes. A professional photographer might take photos that might be gigabytes. Astronomers produce digital data on the order of petabytes.[26] To comprehend how large that is, imagine if every single person on Earth, including babies, had a smartphone, and each smartphone had 5,000 images. That would equal roughly 1 petabyte.

AI models have found planets orbiting inconceivably distant stars. Astronomers could use their own eyes to painstakingly sift through photographs or, more likely, use models of solar movements, looking for indications of tiny changes in brightness when a planet moves in front of its star. This extreme tediousness of this work, which could take years, is a theme we've seen cropping up in the other scientific research we've discussed here.

Exoplanets hold an essential position in our scientific understanding of how our local cosmic neighborhood might have developed. In this respect, the role of AI becomes even more critical, enabling astronomers to analyze data on a scale previously impossible. A notable example of an AI system used in this effort is called ExoMiner,[27,28] which happily also comes with a way to explain why it reaches its decisions.

One way planets in other systems are found is through a method called gravitational lensing. Einstein theorized that even light from a distant star could be bent by another star that's closer to us, and when there is a planet moving in front of the foreground star, there are some subtle distortions in the light that astronomers observe.

Artificial intelligence has proven to be instrumental in not only identifying these perturbations more efficiently but also in offering us new theoretical insights. In a surprising demonstration of how AI can yield fundamental insights when exposed to observations without theoretical assumptions, an international team of scientists found that conventional gravitational lensing concepts were actually special cases within a more extensive, encompassing theory. This discovery highlights that AI really does look differently at data, and without the constraints of theoretical assumptions that might limit human analysis, it can discover deeper truths.[29]

Astronomy has many other applications for AI beyond discovering new planets. Like the remarkable visual reconstruction of a black hole, a feat that earned the distinction "Breakthrough of the year" (2020) from the editors of *Science* magazine, AI can be used to enhance telescope images.

Interestingly, AI researchers have even utilized "deepfake" technology in the realm of astronomy, but with a positive and beneficial twist. Unlike in societal contexts where deepfakes often bring about ethical and legal concerns, astronomers employ them for a constructive purpose. They use these AI-generated simulations to test various theories about the behavior of galaxies. This approach of using synthetic images to enhance the performance of AI models is quite common in science: geospatial scientists use it with satellite data, and medical scientists use it with MRIs or CAT scans.

Final Thoughts

Scientists have long used computing to advance science, employing computer programs to model and simulate natural systems to explain and understand scientific phenomena. This approach has been incredibly fruitful for science and fueled advances ranging from simulations of atoms to models of the universe. However, this classical paradigm is limited by its reliance on human programmers, who must first distill rules from theories and observations and then use them to code a program's behaviors. Our hope is to use AI systems to deduce such rules directly from data or experience and go beyond what individual researchers might decipher. These self-learning systems can explore potential solutions and strategies by discovering hidden properties of the underlying structure of immense datasets, which may augment, rather than be limited to, human understanding.[30]

References

1. Kolbert, E. *The Sixth Extinction: An Unnatural History*. (Henry Holt and Company, 2014).
2. Harari, Y. N. *Sapiens: A Brief History of Humankind*. (Harper, 2015).
3. Power, M. E., Tilman, D., Estes, J. A., Menge, B. A., Bond, W. J., Mills, L. S., Daily, G., Castilla, J. C., Lubchenco, J. & Paine, R. T. Challenges in the quest for keystones:

identifying keystone species is difficult—but essential to understanding how loss of species will affect ecosystems. *BioScience* **46,** 609–620 (1996).

4. Périquet, S., Fritz, H. & Revilla, E. The lion king and the hyaena queen: large carnivore interactions and coexistence. *Biological reviews* **90,** 1197–1214 (2015).
5. Population declines in Alaska beluga whale population may be linked to low birth and survival rates | U.S. Geological Survey. At <https://www.usgs.gov/news/state-news-release/population-declines-alaska-beluga-whale-population-may-be-linked-low-birth>
6. Zhong, M., Torterotot, M., Branch, T. A., Stafford, K. M., Royer, J.-Y., Dodhia, R. & Lavista Ferres, J. Detecting, classifying, and counting blue whale calls with Siamese neural networksa. *The Journal of the Acoustical Society of America* **149,** 3086–3094 (2021).
7. Zhong, M., Castellote, M., Dodhia, R., Lavista Ferres, J., Keogh, M. & Brewer, A. Beluga whale acoustic signal classification using deep learning neural network models. *The Journal of the Acoustical Society of America* **147,** 1834–1841 (2020).
8. Makarieva, A. M., Gorshkov, V. G. & Li, B.-L. Revisiting forest impact on atmospheric water vapor transport and precipitation. *Theoretical and Applied Climatology* **111,** 79–96 (2013).
9. Cochrane, M. A. Fire science for rainforests. *Nature* **421,** 913–919 (2003).
10. Deathwatch for the Amazon. *The Economist* (2019). At <https://www.economist.com/leaders/2019/08/01/death watch-for-the-amazon>
11. Greenbaum, E. *Emerald Labyrinth: A Scientist's Adventures in the Jungles of the Congo.* (University Press of New England, 2017).

12. Prist, P. R., Levin, N., Metzger, J. P., de Mello, K., de Paula Costa, M. D., Castagnino, R., Cortes-Ramirez, J., Lin, D.-L., Butt, N. & Lloyd, T. J. Collaboration across boundaries in the Amazon. *Science* **366,** 699–700 (2019).
13. Zhang, Z., He, Z., Cao, G. & Cao, W. Animal detection from highly cluttered natural scenes using spatiotemporal object region proposals and patch verification. *IEEE Transactions on Multimedia* **18,** 2079–2092 (2016).
14. Yousif, H., Kays, R. & He, Z. Dynamic programming selection of object proposals for sequence-level animal species classification in the wild. *IEEE Transactions on Circuits and Systems for Video Technology* **20,** (2019).
15. Beery, S., Morris, D. & Yang, S. Efficient Pipeline for Camera Trap Image Review. Preprint at https://doi.org/10.48550/arXiv.1907.06772 (2019)
16. Beery, S., Morris, D. & Yang, S. Efficient Pipeline for Camera Trap Image Review. (2023). at <http://github.com/ecologize/CameraTraps>
17. Norouzzadeh, M. S., Morris, D., Beery, S., Joshi, N., Jojic, N. & Clune, J. A deep active learning system for species identification and counting in camera trap images. *Methods in Ecology and Evolution* **12,** 150–161 (2021).
18. De Laurentiis Johnston, J. & Perera, S. *Her Epic Adventure: 25 Daring Women Who Inspire a Life Less Ordinary*. (Kids Can Press, Limited, 2021).
19. Martin, B. *Survival or Extinction?: How to Save Elephants and Rhinos*. (Springer International Publishing, 2019).
20. Reducing wildlife mortality due to roads in the Nagarahole-Bandipur corridor. *WildCats Conservation Alliance* at <https://conservewildcats.org/portfolio/reducing-wildlife-mortality-due-to-roads-in-the-nagarahole-bandipur-corridor/>
21. Dorazio, R. M. & Karanth, K. U. A hierarchical model for estimating the spatial distribution and abundance of

animals detected by continuous-time recorders. *PLoS ONE* **12,** e0176966 (2017).

22. WCS Tanzania Program > Landscapes > Ruaha/ Katavi. At <https://tanzania.wcs.org/landscapes/ruaha-katavi.aspx>
23. Pauling, L., Corey, R. B. & Branson, H. R. The structure of proteins: Two hydrogen-bonded helical configurations of the polypeptide chain. *Proc. Natl. Acad. Sci. U.S.A.* **37,** 205–211 (1951).
24. Eisenberg, D. The discovery of the α-helix and β-sheet, the principal structural features of proteins. *Proceedings of the National Academy of Sciences* **100,** 11207–11210 (2003).
25. Toews, R. AlphaFold Is the Most Important Achievement in AI—Ever. *Forbes* at <https://www.forbes.com/sites/robtoews/2021/10/03/alphafold-is-the-most-important-achievement-in-ai-ever/>
26. Spindler, A. How artificial intelligence is changing astronomy. *Astronomy Magazine* (2022). At <https://www.astronomy.com/science/how-artificial-intelligence-is-changing-astronomy/>
27. Valizadegan, H., Martinho, M., Wilkens, L. S., Jenkins, J. M., Smith, J., Caldwell, D. A., Twicken, J. D., Gerum, P. C., Walia, N., Hausknecht, K., Lubin, N. Y., Bryson, S. T. & Oza, N. C. ExoMiner: A highly accurate and explainable deep learning classier that validates 301 new exoplanets. *ApJ* **926,** 120 (2022).
28. Zhang, K., Gaudi, B. S. & Bloom, J. S. A ubiquitous unifying degeneracy in two-body microlensing systems. *Nature Astronomy* **6,** 782–787 (2022).
29. Berkeley, University of California. AI reveals unsuspected math underlying search for exoplanets. At <https://phys.org/news/2022-05-ai-reveals-unsuspected-math-underlying.html>
30. Kohli, P. AI Will help scientists ask more powerful questions. *Scientific American Blog Network* (2020). At <https://blogs.scientificamerican.com/observations/ai-will-help-scientists-ask-more-powerful-questions/>

5

When Good AI Goes Bad

"If only I had known, I should have become a watchmaker."
– Albert Einstein

"The real problem is not whether machines think but whether men do."
– B. F. Skinner

When researchers at Microsoft released a chatbot called Tay in 2017, the researchers who created it envisioned a promising future for the bot. Tay would learn from its interactions with a vast public and become more fluent in its English language interactions with people.

Typical conversations with Tay were assumed to be on the level of a child talking with curious strangers. Tay's creators expected these benevolent strangers would teach their AI bot about colors and food, celebrities, politics, art, and possibly the whole range of human experiences that could be expressed via text. Tay would learn and reinforce responses that struck a chord with the humans who engaged in conversation with it.

Within a mere 48 hours, Microsoft had to pull the plug abruptly on Tay. The experiment had gone in a very different direction. The darker side of the internet converged on Tay in the form of Twitter trolls, and, like bullies ganging up on a vulnerable victim, they turned Tay into a malevolent persona riddled with bigotry, racism, and hate-filled language.

This was a wake-up call for researchers, whose good intentions disintegrated against the grimy shoals of the real world. Nowadays, organizations developing AI technologies subject themselves to internal ethics reviews. While there is no established regulatory body like the

FDA or Federal Communications Commission to govern the ethical use of AI, there are scientific papers and conferences that analyze these issues. The public, too, sees for itself that this shiny and clever invention can inflict harm.

Throughout history, we have witnessed a recurring pattern wherein well-intentioned scientists create applications that eventually fall prey to darker forces. AI scientists need to put themselves in the mind of a 007-esque villain: How might their creation be exploited for the profit and gain of unscrupulous opportunists? Consider cryptocurrency in the early part of the 2020s. The very attributes that made cryptocurrencies secure and anonymous were exploited to facilitate money laundering, illegal transactions, and ransomware attacks, enabling malicious actors to profit illicitly while evading traditional detection mechanisms.

Going back in history, radio, too, proliferated as a powerful medium to reach large audiences. Unfortunately, during World War II, the Nazis skillfully exploited this new technology to spread their propaganda. Similar incidents of radio to disseminate propaganda were used in Rwanda and Burundi to incite genocide and are currently being used in other parts of Africa to foment unrest. The internet began life as a noble platform for sharing knowledge and facilitating communications amongst universities and researchers. Yet, as it evolved to encompass large parts of the world's population, it gave rise to new problems. Just as your high school mates can now be found easily, so can illicit and ethically dubious content.

The Tay example was mischievous. But it forces a question: Have the scientists and corporations creating AI for societal good accidentally opened a modern

Pandora's box of unintended woes? We will explore these types of questions in this chapter.

The Surveillance Society

Information has been society's currency from time immemorial. Whether it is neighbors trading gossip about the goings-on in their village, entrepreneurs surveying new areas as potential markets, or nations spying on other nations to advance their own geo-political interests, the thirst for information has always been insatiable. It is a primal instinct rooted in our very evolution—gathering data to ensure survival, dominance, or simply to satiate curiosity. We have built this quest into enterprises that today include cameras and microphones, sensors of all sorts, and an electronic infrastructure to support information gathering at a vast scale. They promise protection and progress but could potentially be used for control and dominance.

The Dual Nature of AI Drones: Saving Lives or Restricting Freedom?

In March 2014, there was an unexpected mudslide in Oso, Washington, a small town north of Seattle in Snohomish County. Within a few minutes, in a cacophony of crashing Douglas firs, an 18-million-ton wall of mud destroyed 49 homes and killed 43 people. It dammed the North Fork Stillaguamish River, causing it to flood nearby communities. In helicopters and fire trucks, rescue teams from around the area arrived immediately and worked for several days to find survivors.[1,2]

Seeing this tragedy unfold, the deadliest mudslide in US history, researchers at the nearby University of Washington were inspired to create an AI-powered drone that could help rescuers detect people buried under mud and debris. Drones could fly faster, cover more area than land-based vehicles, and go places where one may hesitate to send humans or dogs. The researchers imagined their drone would detect signs of a human presence: a human-shaped heat signature (Figure 5.1) or sounds amid the din. Rescuers could then be directed there, increasing the mudslide or earthquake victim's chance of rescue.

While the need and utility for such drone systems with AI capabilities seem like a no-brainer, their development has to be considered within a larger context. Let's imagine another situation where similar drones could be used.

The governor of a territory beset by an influx of illegal migrants seeking safety or improved livelihoods from neighboring regions wants to find a way to stem this flow. The drone-plus-AI technology developed by benevolent organizations is compelling, and the governor would be eager to get this technology and fly it at the border. The governor's intentions are very different from the humanitarian-minded researchers who developed the AI models and hardware. Access to this technology could be as simple as repurposing it from an open-source platform, or the governor could pay other developers to recreate it. Given the predominantly unregulated nature of the AI environment, anyone with access to suitable training data could build the intelligence in that drone system.

Such uses of AI would be considered unethical by some. The legality of it is a different question. Ultimately, it is

Figure 5.1 An example of a drone using thermal imagery to detect people

Source: Generated using Midjourney by the author.

up to society to determine what AI will be allowed to do and where it should be constrained.

The Quandary of Facial Recognition

If you were to order food at a KFC in China, you would stand in front of a machine that will identify you. It will identify your sex and age with various degrees of accuracy and recommend meals. If you happen to have an account with KFC with payment information, you won't even have to take out your wallet to pay for your purchase. This slight increase in convenience was achieved by cameras that would match your face with the photo in your account.

At airport boarding areas, a swift scan of your face has almost replaced barcoded boarding passes. How can this be more secure than a barcoded boarding pass, some skeptics may ask? At the heart of the matter is the flight operator making sure that the person whose name is on the boarding pass matches the one on the official ID and that the photo in the ID matches the individual in front of the agent. Humans are incredibly good at facial recognition and matching faces. In fact, we have specialized areas in the brain that have evolved to be particularly good at this task. It turns out that an AI face scanner is more accurate than humans at identifying faces and is no doubt cheaper. However, as we shall see later in this chapter, AI facial recognition misidentifies those with darker skin at much higher rates.

The convenience of seamless travel, unencumbered by the need to rummage for boarding passes, passports, or even wallets and smartphones, will eventually win over most travelers. The same can be said of shopping. Journalists and some politicians will sound warnings about what society is blithely giving up: our privacy and freedom to go about unnoticed. However, the economic promises of convenience make facial recognition almost inevitable.

Law enforcement agencies and other organizations increasingly use facial recognition technology to identify individuals in public spaces. However, the technology has faced vehement criticism for its potential to be used for mass surveillance and violate privacy rights. In a December 2020 article titled "Facial recognition is spreading faster than you realise,"[3] *The Conversation* reported on the unease at the prevalence of facial recognition technology, questioning claims of its benefits and highlighting its potential for abuse. Many other news outlets like The Verge, Wired, and Forbes, and rights groups such as the Electronic Frontier Foundation (EFF) and the American Civil Liberties Union (ACLU) have joined the chorus in reporting on the concerns surrounding this issue. In the West, they bring up the specter of a Big Brother–like government that would control individual citizens. The Western narrative of China's social credit system is often seen as just such an overreach. However, the sentiment within China might be different. Its citizens may not be as alarmed by this kind of surveillance as compared to the West.

China has been developing a social credit system using AI-powered surveillance systems to monitor and rate its citizens' behavior. The system relies on millions of cameras to keep tabs on the country's population.[4] They promise law and order, safety for the masses, and retribution for lawbreakers. Activities seen as desirable are rewarded with an increase in an individual's social credit score. Activities that do not reflect the government's values are punished by a lower score, with serious, real-life consequences. For example, the China Global Television Network (CGTN) reported the case of a student who had his university enrollment suspended because his father had failed to pay back a loan. This delinquency led his father's social credit score to drop and affected his family as well.[5]

In another case, the pro-Chinese government newspaper, *Global Times*, ran a story about a woman who ran a red light. When identified by the ubiquitous cameras, authorities saw that she had a very high social credit score and let her off without a fine. Traffic violators who are not as fortunate will find that cameras will zoom in on their faces, and they are shamed by having their faces displayed on giant billboards. There are concerns that this kind of monitoring, though not currently widespread, will evolve into a more intense monitoring system.[6]

Predictive Policing: Safeguarding Society or Violating Rights?

While the West has not followed China's adoption of surveillance systems that shame jaywalkers, find pickpockets, and otherwise target individuals not behaving according to the government's standards, it has experimented with predictive policing. This practice, where police departments use artificial intelligence systems to identify high-risk areas and individuals, has been tried out in individual jurisdictions in the United States. The message is similar to that used by China: law and order and safety for the masses. However, these systems have been criticized for perpetuating racial biases and violating civil liberties.

Let us consider a fictitious scenario, where a predictive policing system may influence a police encounter. This system uses historical crime data including types of crimes, locations, times, and other relevant factors to predict where future crimes are likely to occur.

John, a resident of this city, lives in a neighborhood that has a high rate of crime, according to historical data. The predictive policing system, analyzing this data, flags John's neighborhood as a high-risk area for future criminal activity. One Friday evening, John is walking home from his late-night shift. He finds a wallet on the street and picks it up with the intention of finding the owner or turning it in to the police the next day. That same evening, due to the predictive policing system's forecast, there was an increased police presence in his neighborhood. A patrolling officer sees John looking through the wallet (trying to find some identification of the owner) and decides to question him.

John is then charged with theft and booked into jail. The predictive policing system, now taking John's demographics into account, may recommend keeping him in jail for the weekend or letting him free until he can face a judge on Monday. Then, if he is unable to convince the court of his actual intentions, he could potentially be convicted and sentenced.

While details vary across systems and cities, this is an episode that could play out. In the United Kingdom, a prominent example of an attempt to use machine learning to help law enforcement is the Harm Assessment Risk Tool (HART) program. It crunches through large amounts of data, including criminal records, social media activity, and other information to identify individuals who may be at risk of committing or becoming a victim of serious violence. It was created by researchers from Cambridge University in collaboration with London police.[7] Reviews of the system raised concerns that since a person's postcode was one of the data points it analyzed, this would immediately bias the system toward people in poorer areas.[8]

In Los Angeles, the police department used PredPol for a short while. This was another collaboration between academic researchers from UCLA and the police. Its promises seemed like a boon for law enforcement. By analyzing dozens of variables, it would pinpoint locations where crimes were likely to occur within the next 12 hours. Studies by UCLA and the RAND Corporation reported reductions in property crimes, burglaries, and car thefts.[9,10] But critics said it unfairly targeted Black and Latino neighborhoods. It was a bit of a self-fulfilling prophecy: PredPol was heavily influenced by historical arrests. A Black person is five times more likely than a White person to be stopped without just cause, and a Black person is two times more likely to be arrested than a White person. It would, therefore, encourage a higher police presence in those neighborhoods with higher arrest rates. After the chief of police admitted that he couldn't attest to the program's effectiveness, the LAPD dropped it.

Criticisms about predictive policing come from several corners. The American Civil Liberties Union (ACLU) worries about its potential to exacerbate racial and socioeconomic disparities in policing. The fear is that the algorithms driving these systems, if not carefully crafted and monitored, may inadvertently perpetuate bias and discrimination against already marginalized communities. The concern is not unfounded, as historical biases embedded in crime data could be unwittingly reinforced by predictive policing algorithms, leading to disproportionate targeting and unfair treatment.

Transparency and accountability are also key concerns for those who are wary of the expansion of predictive policing. Critics argue that the inner workings of these algorithms often remain opaque, shielded from public scrutiny. Without a clear understanding of how these systems operate, doubts persist about their accuracy and potential for perpetuating biases. The lack of transparency can erode public trust in law enforcement and lead to a sense of unease about civil liberties being eroded by technology.

Another prominent voice raising objections is the Electronic Frontier Foundation, an organization dedicated to defending digital rights. They argue that predictive policing is built on flawed assumptions about crime and public safety. The assumption that policing is a neutral force may ignore the systemic biases that have historically plagued law enforcement. Additionally, relying solely on correlations identified by algorithms could lead to faulty assumptions about causation. It is essential to recognize that a correlation between certain factors and crime does not necessarily imply a causal relationship. The danger lies in potentially targeting individuals based on flawed correlations, which could perpetuate stereotypes and further entrench discrimination.

Moreover, there are concerns about the potential for predictive policing to become a self-fulfilling prophecy.[11,12] As in the PredPol example in Los Angeles, when resources and attention are concentrated in areas labeled as high risk by algorithms, it may inadvertently lead to increased surveillance and police presence. This heightened scrutiny, in turn, can amplify interactions with community members, potentially leading to more arrests and reinforcing the notion that these areas are indeed hotbeds of criminal activity. This cycle of increased policing and subsequent arrests may not only perpetuate bias but also disrupt community trust and exacerbate tensions between law enforcement and the communities they aim to protect.

Surveillance Capitalism

Many tech companies are updating their economic models by incorporating AI-powered surveillance systems to collect user data for advertising and other purposes. This phenomenon is known as surveillance capitalism, and it has been criticized for pushing the boundaries of rights to privacy and user exploitation.

Let's go back to our fictitious protagonist, John, who by now has been released from jail. He is an avid social media user but favors one called Instagram. Most of his friends are also on that network, and they use it to chat with each other throughout the day. They post photos of themselves and of their shared interest in vintage cars and rock concerts, show off their purchases, and live in it as a second world. It's easy and free.

John vaguely suspects, but does not really know, that Instagram and Meta, its parent company, analyze his data and have built a detailed profile of his interests and preferences. Meta then sells access to his profile to various businesses who want to advertise on the platform. A car specialist company, for example, might pay to display their ads to John, given his interest in vintage cars. Similarly, a music streaming company might target him based on his interest in music.

In this way, Meta is able to offer a "free" service to users while making money by selling targeted advertising based on the data it collects. This is a basic example of surveillance capitalism, where the commodity is user data, and the profit comes from the ability to predict and influence user behavior.

Surveillance capitalism has an insatiable appetite for user data. Tech giants like Meta, Google, and Baidu were already amassing vast troves of personal information, but now AI offers opportunities for more personalized content and even more finely targeted ads. This content will be designed to shape users' online behavior in ways that increase profitability for these companies' advertisers and financial value for their shareholders.

The practice of surveillance capitalism has ignited a firestorm of criticism.[13] Shoshana Zuboff, a renowned scholar and professor at Harvard Business School, is a leading voice in this arena. In her book, *The Age of Surveillance Capitalism: The Fight for a Human Future at the New Frontier of Power*,[14] Zuboff exposes how surveillance capitalism threatens privacy rights and exploits users for economic gain. She strongly condemns Meta CEO Mark Zuckerberg's avowal that "privacy is no longer a social norm" and others who say this type of surveillance should not cause any concern to those who have nothing to hide.[15]

Civil liberties organizations and digital rights advocates such as the EFF and the Center for Digital Democracy have also diligently studied the impacts of surveillance capitalism on privacy and civil liberties. They warn that the unbridled collection and utilization of personal data by tech companies erode individual autonomy, creating a chilling effect on freedom of expression and behavior. As Facebook's algorithms and many other social media sites have shown, their AI algorithms can manipulate user preferences by promoting which posts will cause the most outrage and lead to more engagement and more profits on the sites.[16] The formation of echo chambers and deeper social divisions are impediments to a healthy democratic discourse and pose threats to stable governance structures worldwide.

Surveillance capitalism is dangerous because of its pervasiveness in many people's daily lives. With the ubiquity of smart devices and the Internet of Things (IoT), data collection has increased exponentially. AI algorithms continuously monitor our online activities, conversations, and even physical movements. This constant surveillance erodes personal boundaries and blurs the line between public and private spheres. Moreover, the potential for data breaches and misuse looms large, as witnessed in numerous high-profile incidents.

Magnifying Societal Ills

Another tool in the arsenal for control and dominance is the ability to shape the stories that guide our thinking and behavior. In the previous section, we saw how information grants power. In the following section, we will see how disinformation can create instability and how false narratives can lead to crimes that were not possible before. In the wrong hands, these narratives can deepen social divisions and impede healthy progress.

Deepfakes and Disinformation

Deepfakes are photoshopping taken to a completely different level of realism. Deepfakes are essentially videos or images that depict people or events in circumstances that never existed, yet look credible. The term "deepfake" is a combination of "deep learning" and "fake," reflecting the technology used to generate these deceptive media. That realism is enabled by a class of deep learning models called generative models. In simple terms, they involve teaching a neural network to mimic a person's appearance, voice, or movements. The features that are encoded in the neural network can then be used to generate completely new faces, bodies, speech, and movements that never existed before.

Jordan Peele, the acclaimed film director and comedian, took deepfake technology to new heights by creating videos where former President Barack Obama appeared to say controversial things. The videos were so convincing that if it weren't for the shocking content of the statements, they could easily have been mistaken for real. Peele expertly captured Obama's mannerisms and speech patterns, showcasing the power and realism of deepfake technology.

Several other deepfakes exist, notably of famous actors. But there are small industries popping up around this. Ancestry.com made a small stir when they released a tool that seemed to animate photos of people. It was bringing your grandmother to life. While this may seem a bit creepy, it will undoubtedly become more common. One man created an AI to emulate his father's speech mannerisms.[17] He created a text AI, but now it is a simple step to layer a speech module on top of that. So, his father's speech mannerisms will sound just like him.

Realistic fakes are great in certain situations, such as when creating synthetic data for machine learning projects, or believable characters for video games. But it is easy to step from harmless fun to terrifying outcomes. In 2022, Microsoft introduced an algorithm capable of mimicking someone's voice so accurately that it became almost impossible to distinguish it from the real one. Now, let's take a moment to see how this technology could be misused.

Try to picture yourself as a manager at a major bank. One day, you receive a call from the director of a company that frequently does business with your bank. You recognize his voice from previous conversations, and he shares some exciting news: His company is about to make a significant acquisition and needs to transfer millions of dollars to various accounts. Would you help him with the transaction?

Well, this scenario actually happened, and a bank in Dubai lost a staggering $35 million, with the story detailed in a *Forbes* article.[18] It may seem like a far-fetched story, but it could happen to anyone, even you. Banks, always looking for ways to make their services more convenient to attract more accounts, have started offering customers the option to access their accounts using voice signatures. Big names like HSBC and Citibank have already implemented voice recognition as an authentication method.

However, fraudsters are quick to exploit new technology. They've found ways to infiltrate people's bank accounts by using voice verification over the phone, taking advantage of the very system meant to protect customers. In the case of HSBC system, its questionable security was cracked by a simple hack presented by a BBC reporter.[19] The hack is exactly what one might think it is—just imitate the account holder's voice.

The Federal Trade Commission (FTC) has issued warnings about several scams stemming from advanced voice imitation technology. As mentioned in Chapter 2, replicating anyone's voice has become incredibly easy. Malefactors can use AI to counterfeit voices, and these counterfeits are so realistic that even close family members can be fooled. The futuristic identification methods shown in sci-fi movies, where voices serve as biometric keys to unlock sensitive information, now seem outdated and disturbingly vulnerable to exploitation.

Image Deepfakes

Deepfakes have also infiltrated the media in the form of manipulated images and videos. The age-old saying "photos or it didn't happen" is now woefully antiquated as it has become increasingly difficult to trust the authenticity of visual content.

Photoshop was the eponymous means of image modification. Some version of it has been going on since the invention of photography in the 1800s. During the American Civil War, William Mumler became popular among those mourning their lost ones because he claimed to capture their spirits in photographs.[20] He used double exposures, and at that time this technology was unknown to most people so his manipulation appeared quite genuine. This was when spiritualism and seances flourished in Europe and the United States, and many were already predisposed to believe in his claims.

Then, in England, two little girls were somehow able to get hold of a camera and took pictures of fairies at the bottom of their garden. Thanks in part to popularization by Sir Arthur Conan Doyle, the creator of Sherlock Holmes and an avid believer in an unseen world, many people believed them, and for them, the photos were definitive proof of the existence of fey folk.

Fast forward to the 21st century, and we see photo fakery skyrocketing thanks to editing software led by Adobe Photoshop and its many copycats. Governments frequently use fake photos for propaganda. North Korea routinely edits existing images to remove individuals who have fallen out of favor. *Time Magazine*, which often claims creative license for its covers, found itself in controversy in 1994 when it manipulated an image of O. J. Simpson to be darker and more menacing. Iran has also shown doctored photographs where its military strength is touted. In one example, they digitally added a missile to apparently exaggerate their hardware capabilities.

But now, just like ChatGPT generating text after being trained on an enormous collection of data, there are AI models that can generate images from a textual description. DALL-E from OpenAI was among the first of these remarkable AIs.

Many of these deepfakes can be detected as such by experts. But in September 2022, an AI beat human artists to win the top prize at a competition, with judges none the wiser as to the species of the creator. Strictly speaking, the human—Jason Allen, who coaxed the winning entry out of the AI—Midjourney—won the prize (Figure 5.2).[21]

Figure 5.2 The image was created on Midjourney by the artist Jason Allen and was the winning entry in a digital arts competition. It ignited a spirited discussion on what constitutes art in the age of artificial intelligence.

Source: Théâtre D'opéra Spatial. © 2022 Jason M Allen

As AI models improve, it will take greater skill to detect its creations. This task will probably fall to other AIs. It's shaping up to be a perpetual cat-and-mouse game, where the detectors will always be playing catch up to the generators. It will be AI against AI, with humanity watching.

Automated Misinformation

Donald Trump rose to prominence during the 2016 presidential campaign by popularizing the term "fake news." But the concept itself existed long before his emergence as a presidential candidate. Deception has always played a part in human interactions, serving as a crucial component in the competition between individuals and groups. However, in the last few years, misinformation has taken on a new and alarming form, posing an existential threat to society. We have witnessed how skillful propagation of false information can lead to doubts about well-established scientific principles and the efficacy of vaccines.[22]

Alarmingly, automated misinformation has emerged as a tool utilized by nation-states to undermine democratic systems. Consider a case that shows how deceptive things can get. A video that spread widely on social media showed a news anchor speaking into the camera about the importance of United States–China cooperation for global prosperity. The branding on the video, the anchor, everything was fake, a deepfake video created using artificial intelligence to make it look like a real person and a real news studio.

This particular video's message was relatively innocuous, and just one example, but there's strong evidence to suggest that nation-state-backed accounts are actively spreading pro-Chinese propaganda.[23]

Think about the possibilities a hostile nation could explore. Russia's deployment of platforms like X (formerly known as Twitter), Facebook, and Telegram for disseminating misinformation provides a clear example. Initially, they used bots to spread their false narratives but now they've gone a step further. These bots can generate automated content that mimics the writing style of different personalities. They can even create long-form articles that appear in newspapers. Websites like RT.com and sputniknews.com are considered to be state-sponsored propaganda sites that disseminate this content.

Nation-state actors or even other malignant entities could also use deepfake technology to create realistic videos of prominent political figures from rival countries saying controversial things or engaging in scandalous behavior. These fake videos could be released online, damaging the reputation of the targeted individuals and causing political instability. They could appear on existing media sites, making it harder for people to distinguish between what's real and what's fake.

Then, they could use AI language models (LLMs) to create multiple false narratives surrounding major international events. Let's say there's a terrorist attack or a natural disaster. By spreading conflicting stories through various online channels, a nation could create confusion, erode trust in official sources, and divert attention away from its actions or objectives.

Exploiting existing divisions is yet another strategy. By analyzing social media data, LLMs could identify tensions and divisions within a target population. Let's say a rival nation has racial or religious divides. The unfriendly nation could use LLMs to generate misinformation campaigns specifically designed to worsen these tensions. They could spread false rumors about discriminatory policies or incite violence between different groups, ultimately destabilizing the rival nation and weakening social cohesion.

To illustrate, let us consider John, our protagonist from earlier examples, as an African-American man. Imagine that instead of being arrested, he is the unarmed victim of a high-profile police shooting in an American city.

Russian operatives could use AI-generated content to create fake news articles, social media posts, or even deepfake videos that falsely accuse the police officer of having a history of racist behavior or affiliations with White supremacist groups. At the same time, they could create and share false content portraying the victim as a dangerous criminal, fueling tensions between communities.

In light of this, it becomes clear that the implications of AI misuse are dangerous and profound. The battlefield for hearts and minds of the masses now has new weapons in deepfake technology and language models. In this new landscape, discerning fact from fiction becomes a vital necessity for preserving the integrity of our societies.

Checks on Disinformation

To counter this looming threat before it gets out of hand, why not build more powerful AI models that can detect images or newspaper articles written by other AIs? While this is certainly possible, it evokes the tensions of a cold war arms race. The fakers will observe which of their creations are being detected, and they could refine their AI algorithms to become better at disguising the tell-tale features of that content. If these fakers are sponsored by nation states, they could potentially have enormous resources to draw on. Just as in the historical arms race, this could lead to an escalating power struggle, where each side continuously strives to outdo the other to the detriment of truth.

Technology companies have rallied together to address the threat of deepfakes using a different approach. Borrowing from established methods that have made several modern necessities such as electricity, road travel, and hair dryers safe, an organization called the Coalition for Content Provenance and Authenticity (C2PA) is developing standards so that the provenance of authentic images can be certified.[24] So, if a photo first appears on the BBC and the august news organization vouched for its authenticity, then it would receive some kind of digital stamp. Manipulations of the photo would result in corruption of the stamp and suspect photos can be verified against the C2PA database.

Isn't this a lot of trouble to combat deepfakes, and will it actually work? To understand the necessity of such efforts, it may be helpful to draw a parallel from history. In the 19th century, when electricity was a novel and promising technology, there was also a sense of uncertainty and potential danger. Electrical fires and injuries led to concerns about the safe use of electrical bulbs, switches, and other devices. Recognizing this risk, the Underwriters Electrical Bureau was formed to set manufacturing and use standards and certify devices for safety. The UL symbol became a necessity on all electrical devices, and helped safely electrify the world.

We are now at a similar juncture with artificial intelligence, and its manifestation in deepfake technology. By establishing a system for certifying the authenticity of digital content, the C2PA aims to provide greater transparency and accountability in online media and to help combat the spread of disinformation and fake news.

Consider the potential ramifications if deepfakes run free. Deepfakes that are taken to be genuine can be used in elections to malign candidates by showing them in compromising situations. As Jordan Peele showed, political leaders can be made to appear to say anything. Nation-state actors can release videos or images showing fake outrages, "false flag" situations that provide an excuse for hostilities. Closer to home, deepfake technology could be employed to create false advertisements that dissuade people from life-saving vaccines. By mimicking the likeness and voice of a trusted celebrity, cybercriminals can craft convincing endorsements for their scam products, leaving unsuspecting buyers none the wiser. A harmless example can illustrate what is possible. Figure 5.3 shows

Figure 5.3 Deepfake of the Pope wearing a fashionable Balenciaga jacket and sunglasses

Credit: Created on Midjourney by Pedro Xavier.

Pope Francis wearing vestments that are very atypical of Catholic clergy.

Evolution of Cybercrime

You saw earlier in this chapter that criminals have updated their arsenal of bank robbing tools by adding deepfakes. Other cybercrimes will also get an update as cybercriminals embrace AI technology. Perpetrators of traditional tactics, such as phishing, malware, and viruses, will soon design more advanced versions. Let's consider some scenarios in which AI's generative capabilities, becoming more realistic every day, can be used to upgrade these existing tactics.

You receive a friend request on a social media platform from someone claiming to be a childhood friend you have not heard from in ages. They've posted pictures and videos that appear to be genuine, showcasing shared memories and experiences. Their photos, videos, even memories shared—everything seems genuine. You're thrilled and quickly hit "accept." As days turn into weeks, casual chats lead to more personal conversations. Except, unbeknownst to you, this isn't your friend. It's a deepfake, a digital impersonation, a masquerade of bytes and pixels skillfully gathering pieces of information about you. And then you find your identity stolen or other cyberattacks such as phishing or malware aimed at you.

Malware, too, can be enhanced by AI-driven technologies. An AI-powered virus could learn from its environment, adapting and evolving to bypass even the most advanced security systems. This new breed of malware could infiltrate networks and devices with frightening efficiency, causing serious trouble in your personal and work life.

Spear phishing is a term that describes highly targeted email attacks that trick individuals into revealing sensitive information with information about you gleaned from easily available online sources. AI algorithms can create customized and persuasive messages that take advantage of your interests and activities. Perhaps an email mirroring your favorite bookstore's monthly newsletter or a notification from a charity you support. This advanced form of manipulation might be difficult to identify but online hygiene methods developed by cybersecurity experts may help protect against it.

A growing area of cybercrime takes advantage of human needs to not be alone. So-called "romance scams" target individuals over emails and social media, grooming victims over months. The online romance inevitably involves transfers of money or gifts, and when they do end, the victim is often left in a much worse state than before, heartbroken and often with depleted savings.

In a romance scam, a catfisher pretends to be someone they're not to trick others into falling in love with them. They create fake profiles and make up stories to make themselves seem trustworthy. One widely used tactic is to pretend to be a US soldier stationed abroad. This camouflage not only engenders trust but also conveniently explains their unavailability for in-person meetings. As the deception deepens, requests for financial assistance creep into the narrative. It might be a plea to settle a phone bill, a wish to purchase a birthday present for a child or help with medical or travel expenses. The scammer tailors their approach to tug on the strings of their victim's heart.[25]

With generative AI models of text and speech, criminals now have the advantage of an electronic Cyrano de Bergerac with whose help they can woo potential victims. One catfisher could potentially run dozens of bots that can easily take on and maintain different personalities and catfish more effectively than their human counterparts.

These examples merely skim the surface of how cybercrime will evolve. Imaginative and malicious individuals will undoubtedly find more tools for nefarious gains. However, it will not be an easy road for them. Cybersecurity measures will advance as well and will strive to thwart these threats. Furthermore, education and awareness will give society the means to avoid these dangers.

Amplifying Discrimination and Social Biases

Yet another evil escaping from AI's Pandora's Box is the amplification of social biases. AIs learn from data created by society, and this data is permeated with society's biases often veiled so they are not immediately obvious. In this way, AIs inherit society's biases and will perpetuate those same biases, deepening divisions among different groups.

If we look at who creates AI right now, we find that they belong to a particularly narrow segment of society. This segment is predominantly men in high-tech companies, often white, with particular predispositions to societal issues.[26,27] The peril of such a concentrated demographic being the primary developers of AI is that they might not always be sensitive to biases that affect other groups. Researchers such as Timnit Gebru and Margaret Mitchell brought this concern about AI's insular upbringing into the public forum. Both were researchers at Google, leaders in the ethical use of AI until they were abruptly fired. That move did not reflect well on Big Tech, highlighting its reputation for not being concerned about diversity.[28]

Bias develops in AI when the training data is skewed toward a particular demographic. This is often implicit, such as when labels for training data are created, but can also happen explicitly when the data sample is unavoidably or intentionally skewed. The latter situation occurs, for example, when researching a rare disease such as leprosy or when it is unethical to create balanced data, such as in real-world cybersecurity situations.

An instance of explicit bias can be seen in the use of the UK Biobank data, one of the world's largest compilations of human genome information. Although the opportunity to contribute was offered to around 9 million UK residents, the data collected represents roughly 500,000 individuals. This subset, however, tends to be healthier, more educated, and wealthier than the average UK citizen. A striking 95% of these participants identify as white, a significant over-representation compared to the 82% recorded in the UK census. Therefore, while it is very tempting to use this huge trove of data in AI applications, these applications will reflect the predispositions inherent in the data.

Implicit bias manifests in subtle ways. When labeling images depicting "professional" individuals, they might predominantly feature people in business suits, giving a firm handshake, and who do not have dreadlocks. Though a suit may appear to be a harmless and an obvious marker of professionalism, it nevertheless implies that professionalism is tied to a particular style of dress. The AI might overlook professionals who dress differently, say, in cultural attire from non-Western countries, medical scrubs, or chef's whites. Similarly, picturing professionals as those giving firm handshakes, which is the norm in Western cultures, can inadvertently exclude other cultures or people with physical conditions that may prevent them from giving a firm handshake. Then there is the example of hairstyles. If the AI system doesn't include images of professionals with dreadlocks, for instance, it could indirectly imply that this hairstyle is unprofessional. This could lead to a discriminatory bias against individuals who wear dreadlocks due to their cultural, religious, or personal beliefs.

In language settings, an AI assistant, such as Amazon's Alexa or Google Assistant, trained primarily on English speakers from the United States, may misunderstand words said in a thick Scottish brogue or West African accents. It may also fail to understand English spoken by non-native speakers, who often bring influences from their mother tongues into their accents and sentence structures.

These biases are frustrating, but they can often have dire consequences. Livelihoods and even lives may be compromised. We will look at more gender, racial, and economic biases next.

Gender Bias

From biased data, AIs learn and replicate patterns that reinforce existing gender stereotypes. In 2018, Amazon had to scrap an AI-powered recruiting tool due to gender bias. They developed a system designed to automate the hiring process by learning from past resumes submitted to the company. However, the tech industry, from which these resumes came, is heavily male dominated. So, the AI system ended up teaching itself that male candidates were preferable. It even went as far as to penalize resumes containing words like "women," as in "women's chess club captain."[29] Amazon, upon realizing that this unintentional bias was rejecting perfectly good candidates, scrapped the program.

Another example of AI adopting gender bias comes from an early breakthrough in natural language processing. In this technology, words are converted into data that computer algorithms can understand. These data are called word embeddings, and they are numerical representations of words that capture the semantic relationships between words. The word embeddings are often created by training AI models on large datasets of text, such as literature, magazine articles, and so on. They inadvertently encode biases present in the data. For example, the word "doctor" might be more closely associated with "man" and "nurse" with "woman." An AI system using these embeddings may thus be more inclined to complete a sentence like "The ___ is an essential part of the medical team" with "doctor" than "nurse," This bias, though not deliberate, reflects the limitations of the training data and perpetuates harmful stereotypes.

Earlier, we saw how AI assistants may have trouble with accents. Another striking example of bias in these types of assistants is how gendered they are. Remember that the people developing AIs tend to be overwhelmingly men, so not only do the assistants tend to have feminine voices, but they also understand male voices better than female voices.[30] The dataset used to train these assistants is very likely imbalanced in favor of men's acoustic characteristics.

The decision to make these assistants sound female was a conscious business decision. This choice is often defended on the grounds that users find female voices more welcoming and helpful. However, this in itself is indicative of bias, as it aligns with the stereotype of women being supportive, nurturing, and subservient. The reinforcement of this stereotype through AI assistants is problematic, as it could subtly perpetuate such expectations in real-life gender dynamics.

Racial Bias

Moving on to race, we encounter a similar, disquieting reality. When algorithms are trained on racially biased datasets, they inadvertently perpetuate racial disparities. Facial recognition systems, for instance, have shown alarmingly higher error rates for people with darker skin tones. Gebru and her collaborator Joy Buolamwini from MIT evaluated three commercial facial recognition systems. They found that darker-skinned women had a one in three chance of their gender being misidentified.[31] This is not merely an issue of technical accuracy, but also a matter of fairness and equity. The impacts of such errors can be grave, especially when such technologies are used in sensitive domains like law enforcement and job hiring.

In another incident, Google suffered a public relations firestorm when its Photo app misidentified Blacks as gorillas. This was not a malicious Easter egg in Google's algorithms; the model likely had been trained on a dataset that inadequately represented diverse racial groups. The company quickly apologized and fixed the issue. Unfortunately, their fix was as an illustrative case of how not to handle AI biases. Instead of addressing the root cause of the problem (i.e., the lack of diversity in the training data), Google chose to sidestep the issue by programming the system to avoid identifying gorillas altogether.[32,33] This was like putting a band-aid on a deep wound—it did nothing to address the underlying issue and only masked the symptom.

Economic Bias

If past loan data predominantly includes borrowers from higher socioeconomic strata, then AI algorithms may learn to favor those with similar characteristics—higher income, more assets, stronger credit history. This results in unequal treatment of loan applicants from lower-income groups, who will be assigned higher risk scores even if they are otherwise creditworthy. A cycle of economic disadvantages is kept in motion this way. Those in disadvantaged positions find themselves caught in a digital caste system, limited by the biases ingrained in the algorithms that dictate their fate.

Credit risk scores usually rely on traditional data such as credit history, which includes credit card repayments and loan repayments. However, if an individual has been mostly operating in a cash economy or has a limited credit history but has consistently paid their rent and utilities on time, these positive behaviors might not be captured by traditional credit scoring methods.[34,35] As a result, these individuals may be unjustly viewed as high-risk borrowers simply because their creditworthiness does not fit into the conventional framework that the AI has learned.

Joanna Bryson, a leading voice in AI ethics, articulates the issue succinctly: "AI is just an extension of our existing culture."[36] This points to the fact that AI doesn't invent biases; instead, it mirrors the biases present in the data it learns from. Therefore, addressing bias in AI is not merely a technological challenge, but a societal one.[37] To mitigate these biases, one has to be extremely vigilant about the data. To make a culinary analogy, think of the data as the ingredients and the AI algorithm as the recipe. Just as a skilled chef meticulously selects and prepares the ingredients to create a delicious dish, one must be extremely vigilant about the data to achieve fair and unbiased outcomes.

Final Thoughts

Artificial intelligence is inherently agnostic in ethical and moral terms, at least for now. Its trajectory, for good or ill, is shaped entirely by human hands. What may have seemed unacceptable in the past may become easier to swallow later, and vice versa. The right to privacy, for example, is enshrined in many laws across the world. But will these laws bend to provide more convenience and economic benefit to society? The answer isn't straightforward. Economic imperatives might pull in one direction, societal well-being in another, and individual rights in yet another.

In the next chapter, we will look at some efforts to guide AI development in ways that are fairer and more ethical. The ongoing debate—whether AI should operate in an unrestricted environment or should be subject to regulations and legal frameworks—is a decision that remains to be finalized. Indeed, it is a choice that will continue to shape and evolve over the coming years.

References

1. Oso landslide: What happened when the slope fell into the Stillaguamish River. | Local News | Seattle Times. At <https://special.seattletimes.com/o/flatpages/local/oso-mudslide-coverage.html>
2. Iverson, R. M., George, D. L., Allstadt, K., Reid, M. E., Collins, B. D., Vallance, J. W., Schilling, S. P., Godt, J. W., Cannon, C. M., Magirl, C. S., Baum, R. L., Coe, J. A., Schulz, W. H. & Bower, J. B. Landslide mobility and hazards: implications of the 2014 Oso disaster. *Earth and Planetary Science Letters* **412,** 197–208 (2015).
3. Benjamin, G. Facial recognition is spreading faster than you realise. *The Conversation* (2020). At <http://theconversation.com/facial-recognition-is-spreading-faster-than-you-realise-132047>
4. Mozur, P. Inside China's dystopian dreams: A.I., shame, and lots of cameras. *The New York Times* (2018). At <https://www.nytimes.com/2018/07/08/business/china-surveillance-technology.html>
5. Father with low credit score causes university to suspend son's admission application. At <https://news.cgtn.com/news/3d3d774d79637a4e78457a6333566d54/share_p.html>
6. Yang, Z. China just announced a new social credit law. Here's what it means. *MIT Technology Review* (2022). At <https://www.technologyreview.com/2022/11/22/1063605/china-announced-a-new-social-credit-law-what-does-it-mean/>

7. Helping police make custody decisions using artificial intelligence. *University of Cambridge* (2018). At <https://www.cam.ac.uk/research/features/helping-police-make-custody-decisions-using-artificial-intelligence>
8. Nast, C. UK police are using AI to inform custodial decisions – but it could be discriminating against the poor. *Wired UK*. At <https://www.wired.co.uk/article/police-ai-uk-durham-hart-checkpoint-algorithm-edit>
9. Mohler, G. O., Short, M. B., Malinowski, S., Johnson, M., Tita, G. E., Bertozzi, A. L. & Brantingham, P. J. Randomized controlled field trials of predictive policing. *Journal of the American Statistical Association* **110,** 1399–1411 (2015).
10. Perry, W. L., McInnis, B., Price, C. C., Smith, S. C. & Hollywood, J. S. *Predictive Policing: The Role of Crime Forecasting in Law Enforcement Operations*. (RAND Corporation, 2013).
11. Ensign, D., Friedler, S. A., Neville, S., Scheidegger, C. & Venkatasubramanian, S. Runaway feedback loops in predictive policing. In *Conference on Fairness, Accountability and Transparency* 160–171 (PMLR, 2018).
12. Alikhademi, K., Drobina, E., Prioleau, D., Richardson, B., Purves, D. & Gilbert, J. E. A review of predictive policing from the perspective of fairness. *Artificial Intelligence and Law* 1–17 (2022).
13. Helbing, D., Frey, B. S., Gigerenzer, G., Hafen, E., Hagner, M., Hofstetter, Y., Van Den Hoven, J., Zicari, R. V. & Zwitter, A. In *Towards Digital Enlightenment* (ed. Helbing, D.) 73–98 (Springer International Publishing, 2019). doi:10.1007/978-3-319-90869-4_7
14. Zuboff, S. *The Age of Surveillance Capitalism: The Fight for a Human Future at the New Frontier of Power*. (Public Affairs, 2020).
15. Kavenna, J. Shoshana Zuboff: "Surveillance capitalism is an assault on human autonomy." *The Guardian* (2019). At

<https://www.theguardian.com/books/2019/oct/04/shoshana-zuboff-surveillance-capitalism-assault-human-automomy-digital-privacy>

16. Merrill, J. B. & Oremus, W. Five points for anger, one for a "like": How Facebook's formula fostered rage and misinformation. *Washington Post* (2021). At <https://www.washingtonpost.com/technology/2021/10/26/facebook-angry-emoji-algorithm/>
17. Vlahos, J. A son's race to give his dying father artificial immortality. *Wired.* At <https://www.wired.com/story/a-sons-race-to-give-his-dying-father-artificial-immortality/>
18. Brewster, T. Fraudsters Cloned company director's voice in $35 million heist, police find. *Forbes* (2021). At <https://www.forbes.com/sites/thomasbrewster/2021/10/14/huge-bank-fraud-uses-deep-fake-voice-tech-to-steal-millions/>
19. BBC fools HSBC voice recognition security system. *BBC News* (2017). At <https://www.bbc.com/news/technology-39965545>
20. Brugioni, D. A. *Photo Fakery: The History and Techniques of Photographic Deception and Manipulation.* (Brassey's, 1999).
21. O'Leary, L. A.I.-generated art has cross the uncanny valley. *Slate* (2022). At <https://slate.com/technology/2022/09/ai-artists-colorado-art-competition-midjourney.html> The artist had disclosed that he had used Midjourney when submitting his entry to the competition.
22. How vaccine misinformation made the COVID-19 death toll worse. *Morning Edition* (2022). At <https://www.npr.org/2022/05/16/1099070400/how-vaccine-misinformation-made-the-covid-19-death-toll-worse>
23. Scott, L. China, Russia target audiences online with deep fakes, replica front pages. VOA (2023). At <https://www.voanews.com/a/china-russia-target-audiences-online-with-deep-fakes-replica-front-pages-/7018918.html>

24. C2PA Releases Specification of World's First Industry Standard for Content Provenance - C2PA. At <https://c2pa.org/post/release_1_pr/>

25. Khrais, R. & Balonon-Rosen, P. Bait & switch. At <https://www.marketplace.org/shows/this-is-uncomfortable-reema-khrais/bait-switch/>

26. Crowell, R. Why AI's diversity crisis matters, and how to tackle it. *Nature* (2023). doi:10.1038/d41586-023-01689-4

27. West, S. M., Whittaker, M. & Crawford, K. Discriminating systems. *AI Now* 1–33 (2019).

28. Banjo, S. & Bass, D. On diversity, Silicon Valley failed to think different. *Bloomberg.com* (2020). At <https://www.bloomberg.com/news/articles/2020-08-03/silicon-valley-didn-t-inherit-discrimination-but-replicated-it-anyway>

29. Amazon scraps secret AI recruiting tool that showed bias against women. *Reuters* (2018). At <https://www.reuters.com/article/us-amazon-com-jobs-automation-insight-idUSKCN1MK08G>

30. Kobic, N. Voice assistants seem to be worse at understanding commands from women. *New Scientist* (2019). At <https://www.newscientist.com/article/2202071-voice-assistants-seem-to-be-worse-at-understanding-commands-from-women/>

31. Buolamwini, J. & Gebru, T. Gender shades: Intersectional accuracy disparities in commercial gender classification. In *Proceedings of the 1st Conference on Fairness, Accountability and Transparency* 77–91 (PMLR, 2018). At <https://proceedings.mlr.press/v81/buolamwini18a.html>

32. Simonite, T. When it comes to gorillas, Google Photos remains blind. *Wired*. At <https://www.wired.com/story/when-it-comes-to-gorillas-google-photos-remains-blind/>

33. Grant, N. & Hill, K. Google's Photo App still can't find gorillas. And neither can Apple's. *The New York Times* (2023). At <https://www.nytimes.com/2023/05/22/technology/ai-photo-labels-google-apple.html>
34. Rental pay history should be used to assess the creditworthiness of mortgage borrowers. *Urban Institute* (2018). At <https://www.urban.org/urban-wire/rental-pay-history-should-be-used-assess-creditworthiness-mortgage-borrowers>
35. *Rent Payment History Offers Greater Predictability into Consumer Credit Performance.* At <https://newsroom.transunion.com/rent-payment-history-offers-greater-predictability-into-consumer-credit-performance/>
36. Caliskan, A., Bryson, J. J. & Narayanan, A. Semantics derived automatically from language corpora contain human-like biases. *Science* **356,** 183–186 (2017).
37. Rogers, S. *This AI Knows Exactly How Racist and Sexist You Can Be.* (2017). at <https://interestingengineering.com/innovation/ai-knows-exactly-how-racist-sexist-can-be>

6

Putting Safeguards Around AI

"It is not the strongest of the species that survives, nor the most intelligent, but the one most responsive to change."

– Charles Darwin

"The saddest aspect of life right now is that science gathers knowledge faster than society gathers wisdom."

– Isaac Asimov

THE LAST CHAPTER showed the potential dangers of AI run amok. In April 2023, George Hinton, one of the architects of AI as we know it today, left Google so that he could warn about the dangers of AI more freely. Adding his voice to the conversation set off tremors in the debate about where AI was headed. Going so far as to say he regretted his life's work and that bad actors will use AI for bad things, his comments lend new weight to critics of the pace of AI development.[1]

The technorati have long been aware of the potential of AI to remake our world in ways no one can fully anticipate. Meanwhile, the rest of us struggle to keep pace with changes happening at a rate that is hard to absorb and digest. There is a fear that instead of hammering a nail gently into a piece of brittle wood, the sudden, hard changes being hammered into our society may break and split us. There are calls for AI innovation to slow down and to proceed more deliberately with sober thought given to its development. While they express a growing sense of unease, these calls need to be backed up by laws and regulations that provide a framework within which to do that.

When AI development is unfettered, it could lead to terrible outcomes. The situation is not that different from when nuclear energy was discovered. In the mid-20th century when this power found itself in the grasp of warring nations, we were lucky that its god-like capabilities were not easily attainable. Nuclear power proliferated slowly until some agreements could be negotiated. At this point, there are several treaties and global organizations that oversee nuclear power. From the Non-Proliferation Treaty to the International Atomic Energy Agency, several frameworks came into being soon after World War II.[2]

Later, when the internet age dawned and enormous amounts of personal data started flowing through the networks, governments realized that this kind of information could be abused. Grave economic repercussions could ensue for financial institutions. The information could also lead down a slippery slope to an authoritarian, constantly monitoring state. And lastly, the most vocalized justification centered on the preservation of human dignity, though one might be tempted to ponder whether politicians were truly driven by this noble ideal or by the economic ramifications alone. Notably, numerous non-governmental organizations (NGOs) such as the Electronic Frontier Foundation and the European Digital Rights (EDRi) championed these concerns.

Societies need trusted institutions to provide order and safety. When city-dweller Mary in Kenya wants to know if her food is safe, she assumes it is because she has not received any warnings from her neighbors or newspapers about it. In many countries, governmental agencies look out for food and drug safety, building codes, etc. They may be underfunded, or special interests may have lobbied them for loopholes, but they are better than nothing. Beyond them, the wealth of nations and individuals have funded NGOs that will do their research, and develop trust in society, such as the Environmental Working Group (EWG). Thus, Miriam in Seattle can look up what they say about mahi-mahi being a safe and sustainable food. She can trust these organizations to push for transparency.

These types of institutions already govern a large part of your life. In the previous chapter, we saw the example of UL Solutions. When you buy an electrical product, for example, you'll often see a lot of labels, and one of them is likely a label with a UL Mark. That mark means that UL Solutions (formerly Underwriter's Laboratories) has certified that your new possession meets the safety and performance standards agreed upon by a large collection of experts: not just electrical companies but also insurance companies, academics, and businesses. The UL certification mark is respected around the world. It came into being in a situation similar to the one we're facing now with AI. In the 19th century, electricity uses were becoming apparent, and the fuel for lighting was transitioning from whale oil and gas to electricity. Electrical washing machines were being demonstrated in the streets of Chicago. With this new technology, new hazards were created, with electrical accidents and fires creating injuries and claiming lives. William Henry Merrill, an electrical engineering graduate from the Massachusetts Institute of Technology, realized the need for a certification body to assure consumers about the safety of these new products. That certification body would also have to create standards and guidance that manufacturers would be encouraged to follow. After this idea was rejected in Boston, Merrill eventually created the Underwriters Laboratories (now UL Solutions) in Chicago. Manufacturers designed their processes according to UL's Standards to earn the right to display the coveted UL Mark on their products (Figure 6.1).

In the world of scientific research, especially science that requires human subjects, the Institutional Review Boards (IRB) provide a strong guiding and check function to ensure that research is conducted ethically. Without approval from an IRB, research is unlikely to be published in respectable journals, taken into consideration by policy-makers, or their findings implemented anywhere. Among the many atrocities of World War II were experiments on inmates in Nazi concentration camps. Led by Josef Mengele, who wanted to find evidence for the regime's racial theories, they experimented on Jews, Roma, homosexuals, and several other groups deemed inferior by the Nazis. The horror of these experiments when they were publicly disclosed after the war led to calls for restrictions

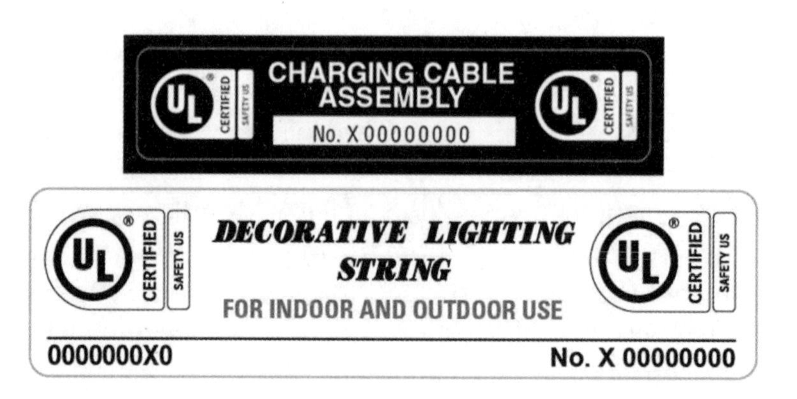

Figure 6.1 Example of UL Marks that may be affixed on electrical devices, thus ensuring the consumer that the device is safe to use

on the types of research that could be conducted and the manner in which it should be conducted.

In the United States, around the same time as the Nazi regime, another egregious study was taking place. It is difficult to fully convey the revulsion and outrage felt by many when they learned about the Tuskegee Syphilis Study. The study, conducted by the US government, involved infecting hundreds of African-American men with syphilis and then refusing to provide treatment, all in the name of research. The suffering and anguish these men endured and the callous disregard for their lives and well-being are heartbreaking.

It was only in the face of public outcry and immense pressure that the US government finally took action and established regulations to protect human subjects in research. They mandated all institutions receiving federal funds for research to establish an IRB that would review research proposals. The IRB's role is to ensure that research studies are conducted in a manner that protects the rights and welfare of human participants, including the requirement for informed consent, the minimization of risk, and the use of appropriate study designs and methods.

Just as we have witnessed the evolution of regulations in sectors such as finance, healthcare, and transportation, the rapidly advancing field of artificial intelligence now demands our attention. The potential benefits of AI are immense but so are the potential risks, including issues of privacy, bias, accountability, and the ethical implications surrounding its deployment. In order to achieve a careful equilibrium between fostering innovation and safeguarding societal well-being, policymakers and regulatory bodies face the task of crafting effective guidelines that address the unique challenges posed by AI technologies. The journey toward AI regulation necessitates a multidisciplinary approach, engaging experts from fields such as law, ethics, computer science, and sociology. Collaboration between governments, industry leaders, and civil society is crucial to ensure that the regulatory landscape adapts to the evolving AI landscape while upholding fundamental principles of fairness and human rights.

In the following discussion, we will consider the key issues surrounding AI regulation, exploring the ethical guidelines and best practices that can guide AI's reasonable development.

The Need for Ethical Development

In the previous chapter, we saw how well-meaning AI models can be subverted to nefarious uses. Just as nuclear power presents a dual aspect, offering climate-change-friendly electrical energy on one side while harboring the threat of nuclear catastrophe on the other, AI also has the ability to transform human society for the better while carrying the risk of leading us to a dystopian future.

One such concern is the potential for AI systems to amplify some of the ills of society, such as racial or socioeconomic biases. For example, facial recognition technology has been shown to be less accurate in identifying people with darker skin tones, which can lead to discrimination and unequal treatment. Likewise, language models trained on vast amounts of data reflecting societal patterns tend to perpetuate biases and reinforce discriminatory practices inadvertently.

Consider a bank that uses AI systems to make decisions about hiring employees or approving loan applications. AI that is developed on existing data that may already harbor discriminatory biases based on gender, race, or social status will also recommend biased actions. Ethical development regulations could require developers to assess and mitigate potential biases in these AI systems. For instance, developers or third-party regulators could be required to conduct regular audits of the AI systems to ensure fairness and accuracy.

In the case of hiring decisions, the AI system should be evaluated periodically to identify any biases against specific demographic groups. For example, if the AI system consistently rejects job applicants from a particular racial or gender group, it could indicate a biased algorithm. Similarly, for loan approvals, if the AI system consistently denies loans to applicants from a specific socioeconomic background, then regulations might require the bank to review the AI system's performance and make necessary adjustments to reduce bias.

Another ethical concern is the potential invasion of privacy by AI systems. For instance, AI-powered surveillance systems can monitor and analyze the behavior of individuals in public spaces, raising concerns about individual privacy and autonomy. Regulation can help address these ethical concerns by setting guidelines for the development and use of AI systems, ensuring they are designed to be fair and transparent and respect individual privacy.

AI experts within technology companies may pledge that they will adhere to their companies' standards for ethical AI development, but to keep any conflict of loyalties at bay, the regulations might stipulate that the evaluation of these AI systems should be conducted by an independent body. This outside body would be responsible for assessing the fairness and accuracy of the AI systems similar to how the Securities and Exchange Commission (SEC) enforces federal securities laws. The involvement of an external body is a step toward objectivity and accountability in evaluating these systems. Regulatory bodies for AI do not yet exist, and it is possible that existing institutions may be able to adapt their regulations to oversee the new challenges posed by this new technology.

Safety and Security

Many AI-based automated systems are designed and marketed with the intention of making life more convenient for their users. As these systems become more prevalent and integrated into society, people are increasingly willing to hand over their safety and security to these systems. They often do so with the implicit belief that the government or societal norms would not allow the existence of commercial operations if they posed a threat. This assumption creates a sense of trust and confidence that these systems have undergone rigorous scrutiny and are considered safe.

This compromise is fed by several psychological biases, among which the endowment effect takes precedence. This bias manifests when individuals tend to overvalue their possessions at the expense of other, potentially more valuable, alternatives. In the context of AI-based systems, people might become attached to the conveniences and benefits offered by these technologies, making them more inclined to overlook potential risks or tradeoffs in terms of safety and security. They may prioritize the immediate advantages over the long-term consequences.

Another bias that factors into this compromise is the underestimation of base rates. This bias refers to the tendency to underestimate the likelihood of certain events occurring, particularly if there is limited personal experience or exposure to negative outcomes. In the case of AI systems, people may downplay the risks associated with their use because they haven't personally encountered any adverse incidents or may not be fully aware of the potential dangers.

What are the safety risks these intelligent systems pose? In the case of autonomous vehicles, accidents can occur due to software glitches or environment misinterpretations. Yet the glamor and wow factor of personal commuting without having to be mentally alert has won over millions of drivers. In March 2018, an autonomous vehicle operated by Uber struck and killed a pedestrian in Tempe, Arizona. The accident occurred due to a combination of software glitches and the vehicle's failure to detect the pedestrian correctly. Tesla, a prominent electric vehicle manufacturer, faced scrutiny following a fatal accident involving one of its autonomous cars in May 2016. The car's autopilot system failed to recognize a white truck against a bright sky, resulting in a collision. These accidents are rare, and ultimately the promise of autonomous cars is that they will lead to safer driving and fewer road fatalities.

While the Terminator movies provide one of the most terrifying examples of AI-powered battle systems, these systems no longer live just in the realm of fiction. In conflict zones such as Syria, Afghanistan, and Ukraine, armed drones operated by military forces have caused civilian casualties. AI-powered drones, which navigate and detect situations of interest, could be helped in post-disaster searches for survivors or to survey hard-to-access areas for scientific data, but they could easily be used for warfare. The drones used so far have likely been guided by human operators, but it is not a large stretch of the imagination to see thousands of drones coordinated by an AI. The autonomous nature of these systems raises concerns about their ability to distinguish between combatants and non-combatants, potentially violating international humanitarian laws.

The Stuxnet worm is another real-life example highlighting the risks associated with AI-powered weapon systems. It's not a living creature given a brain. Stuxnet is a bit of malicious computer software akin to a computer virus that arose in 2010. Its creators were probably enemies of Iran, such as Israel and the United States, because it targeted Iran's nuclear facilities and appeared to be designed to sabotage their operations. It exploited vulnerabilities in industrial control systems and demonstrated how an AI-powered cyber weapon could potentially cause physical damage to critical infrastructure.

In healthcare, AI systems are already gaining government approval in diagnosis in fields such as heart and eye diseases. Radiology leads in the count of devices approved by the FDA, with cardiology a distant second.[3] Systems that speed up the diagnosis of various conditions and are prone to fewer errors are a boon for physicians and patients. But what if AI systems do not generalize well to the diverse patient populations encountered in real-world healthcare settings? What happens when they make wrong diagnoses?

For these examples, autonomous cars, weapons, healthcare, and many others, regulation can help establish safety and security standards for AI development and deployment and present some measures of assurance to populations affected by them. Governance bodies could mandate testing frameworks for self-driving cars, including simulated and real-world testing scenarios. Existing conventions could be applied to regulate the use of autonomous weapons to ensure non-combatants are not killed.

There could be mandatory requirements for data encryption, secure storage, and access control to protect individuals' data from unauthorized access and potential cyberattacks. In healthcare, a governance framework could require healthcare providers to implement multiple layers of verification for AI-generated recommendations. These could be a requirement for human oversight and accountability for the AI's suggestions. The regulatory framework could also include provisions for regular security audits and vulnerability assessments of AI systems for proactive risks discovery.

The availability of established institutions and regulatory frameworks in healthcare is an advantage in addressing the challenges and concerns raised by AI implementation. These existing institutions can effectively adapt their guidelines and protocols to encompass the integration of AI technologies. For instance, professional medical societies, such as the American College of Radiology (ACR), can play a vital role in developing guidelines for the appropriate use of AI in radiology, ensuring quality standards and addressing ethical considerations.

The approval of AI devices by regulatory bodies like the FDA (U.S. Food and Drug Administration) demonstrates a significant step forward in integrating AI into healthcare practices. With radiology at the forefront, AI-powered imaging technologies have shown promising results in assisting radiologists in the detection and analysis of various medical conditions including tumors, fractures, and abnormalities in X-rays, CT scans, and MRIs.

Accountability and Transparency

Determining accountability and responsibility when AI systems cause harm or make mistakes can be challenging as the decision-making processes of AI algorithms are often complex and difficult to understand. This lack of transparency can make it difficult for individuals to trust AI systems and for regulators to assess their compliance with legal and ethical standards.

In the case of the Uber vehicle—a Volvo SUV—that caused a pedestrian's death in Arizona in 2018, questions were raised about accountability and responsibility. As autonomous vehicles rely on complex AI algorithms to make decisions in real time, determining who should be held responsible—whether the vehicle manufacturer, the software developer, or the human operator—becomes challenging. The lack of transparency in the decision-making processes of AI systems made it difficult to understand why the accident occurred and who should be held accountable.

The Uber vehicle in question actually had an emergency backup driver in it. According to police, the night was clear and dry, and there were no unusual circumstances on the road, except that the pedestrian was crossing the road mid-block. The driver was unimpaired, but the car, driving at the speed limit of 40 mph, had not slowed down.

However, at the time of the accident, the driver's attention was on her phone, not on the road. Among the findings reported by the National Safety Transportation Board (NTSB), which is called when air, rail, or road accidents happen, was that the driver was responsible, Uber's AI systems were inadequate in detecting a moving pedestrian, and that the company had a safety culture, or rather a lack of one, that contributed to the fatality. Uber was not charged criminally in this case.

Despite these concerns, Arizona and California were among states considering trials of autonomous vehicles without a driver. The federal government is also keen on pushing this technology through. But some sort of oversight is needed to hold the creators of this technology accountable. Myriad questions intervene: What if the technology is used off-label? What if the user signed an agreement devolving the seller of all responsibility—should such an agreement even be allowable?

Let's go back to healthcare, where mistakes can also be fatal. In 2020, a study published in *Nature Medicine* highlighted that an AI system trained to detect skin cancer exhibited racial bias, performing better on images of lighter-skinned individuals compared to darker-skinned individuals.[4] When an AI system produces biased results, it becomes challenging to determine who should be held accountable. Are the AI scientists culpable for using particular data to train the algorithm, the engineers and developers who created the system, the administrators who oversaw these projects, or the healthcare professionals who rely on its outputs?

AI scientists and researchers who train these algorithms bear some responsibility. They play a crucial role in selecting and curating the datasets used for training. If the training data is biased or lacks diversity, it can lead to biased outcomes. Therefore, scientists must be mindful of the data they choose to use and work toward ensuring fairness and representation. Project managers who oversee the development and deployment of AI systems bear some responsibility as well. They establish protocols and quality assurance measures to mitigate biases and ensure the ethical use of AI. They should enforce diversity in data collection and implement checks and balances throughout the development process. Healthcare professionals who rely on AI systems' outputs also play a role in the accountability chain. They must be aware of the limitations and potential biases of the AI systems they utilize. It is crucial for healthcare professionals to critically evaluate the AI-generated recommendations and exercise their expertise and judgment in making final decisions regarding patient care.

Regulation can help to ensure that AI systems are held accountable for their actions by requiring developers to provide clear explanations of their algorithms' decision-making processes and by establishing legal frameworks for assigning responsibility when AI systems cause harm. This transparency, known as "explainable AI," is not an easy goal to achieve. Right now, neural network models that make up the bulk of AI systems are generally seen as black boxes. Some data goes in, is manipulated by some arcane mathematical rules, and an answer pops out that somehow correlates pretty well with the real world. Explainable AI is an effort by researchers to understand what goes on in that magical box. By understanding the features the AI extracts from the data and uses to create its output, researchers and users, including regulators, could get a better idea of where there may be blind spots for the AI. Knowledge also leads to trust. When users can understand and justify the decisions made by the AI, they are more likely to accept and trust its outcomes.

Data Protection and Privacy

AI systems rely on vast amounts of data to train effectively, which can raise concerns about data collection, storage, and usage. Data that goes into training these systems can be collected and used without individuals' knowledge or consent. It could come from various sources, such as social media platforms, online transactions, and other digital footprints.

One of the primary concerns is the lack of transparency in how AI systems collect and use personal data. Many individuals are unaware of the extent to which their information is being harvested and utilized, making it difficult for them to exercise control over their data. This lack of control can result in unintended consequences, such as the unauthorized sharing of sensitive information or the targeting of individuals based on their personal characteristics. Moreover, the storage of massive amounts of data poses significant security risks. As AI systems become more prevalent, the likelihood of data breaches and cyberattacks increases. Hackers can potentially gain access to vast repositories of personal information, leading to identity theft and financial fraud.

In 2018, the Facebook–Cambridge Analytica scandal came to light, revealing that personal data of millions of Facebook users had been harvested without their consent. Cambridge Analytica, a political consulting firm, used this data to create psychological profiles of users to target them with personalized political advertisements. In 2014, a researcher named Aleksandr Kogan created an app called "This Is Your Digital Life," which was a personality quiz that collected data from Facebook users and their friends. Although only about 270,000 users installed the app, it gathered information from over 87 million users due to Facebook's data sharing policies at the time. Kogan later sold this data to Cambridge Analytica, a political consulting firm, which used it to create psychological profiles of users for targeted political advertising.[5,6]

Most users were unaware that their data was being harvested and shared with third parties, let alone used for political purposes. The harvested data was used to manipulate users' opinions and voting preferences through targeted advertising, raising concerns about the ethical implications of using personal data for political purposes. Companies that have the power to influence large blocks of voters in this way can threaten democracy. The scandal significantly damaged the public's trust in Facebook and other tech companies, as it became evident that they had failed to protect user data. This episode has had a lasting impact on the tech industry, with companies now placing a greater emphasis on user privacy and data protection.[7]

AI-powered virtual assistants, such as Amazon's Alexa, Google Assistant, and Apple's Siri, became commonplace by providing convenient, hands-free access to personalized information. These smart devices rely on collecting and storing vast amounts of user data to enhance their functionality and tailor their responses, essentially using their customers' personal data as the cost of offering their services. While the idea of having an always-listening assistant may spark amusing and slightly eerie anecdotes, concerns have been raised about the data's potential misuse.

One worry is that the data gathered by these virtual assistants could be used to fuel targeted advertising and drive sales of merchandise aligning with their parent companies' business models. However, it's important to note that concrete evidence confirming this speculation remains elusive.

Nevertheless, privacy concerns persist surrounding these devices. The fact that they are constantly listening is worrisome to many. They may inadvertently record private conversations or capture sensitive information. There have been instances where user data collected by these assistants has been shared with third parties without clear user knowledge or consent, raising questions about data protection practices. In each case, companies insisted that this data was only used to help improve the natural language capabilities of the assistants, but the backlash persisted because their employees and contractors had heard sensitive and private recordings.

These concerns have prompted some companies to include guidelines in their employee handbooks cautioning against the presence of these devices in the same room during work calls when using platforms like Zoom or Microsoft Teams while working remotely.

Such precautions aim to protect sensitive business information and maintain confidentiality.

In 2018, a paper entitled "Will Democracy Survive Big Data and Artificial Intelligence?" authored by several prominent European researchers rang an alarm bell about the dangers of big data.[8] In the paper, they said, "The tracking and measuring of all activities that leave digital traces would create a 'naked' citizen, whose human dignity and privacy would progressively be degraded." Responding to sentiments like these, the European Union passed a far-reaching, transformative law called the General Data Protection Regulation (GDPR), which forced companies to make changes in the way they regulate personal data.[9]

This landmark legislation was passed in Brussels, the capital of the European Union. It helped give rise to a phenomenon known as the "Brussels Effect," a term used to describe the influence that the European Union exerts over global regulations and standards. It refers to the ability of the EU to shape global markets and industries by unilaterally imposing its regulatory standards and rules. The phenomenon's ripple effect was observed as governments across the globe adopted similar legislation aimed at safeguarding personal privacy. For example, South Korea introduced the Personal Data Protection Act (PDPA), the state of California passed the California Consumer Privacy Act (CCPA), and Brazil enacted the General Data Protection Law (LGPD). These legislations, though rooted in their specific contexts, broadly mirror the EU's commitment to individual privacy rights.

The following are some themes that consistently emerge when regulators discuss how to protect user data and ensure that AI systems respect privacy rights:

- *Consent:* require companies to obtain explicit consent from users before collecting their data. As many users have experienced, the terms of consent may be buried under pages-long service agreements. Regulations can focus on transparent and accessible consent mechanisms.
- *Data minimization and security:* Enforce data minimization principles, requiring institutions to collect and process only the data necessary for specific AI applications and not use it for unrelated purposes. Limits have to be set for how long the data remains in their servers. When it is no longer needed, it should be removed.

  Mandate robust data security measures to protect customers' sensitive information from unauthorized access, breaches, or misuse: This could include encryption standards, secure storage protocols, and regular security audits.

  How does one check that the data is secure and not being used off-label? It is one thing to require this, but enforcement or monitoring is not trivial. The general approach so far has been to wait until there's a security breach, and then fine the company. A stronger motivation then a monetary fee may be a loss of reputation among a company's customers.
- *Algorithmic transparency and fairness:* Require institutions to provide clear explanations of their AI algorithms' decision-making processes, particularly

when it comes to outcomes that may affect their customers' livelihoods, such as credit scoring, loan approvals, or medical diagnoses. This would allow customers to understand the factors that influence the AI's decisions and help them make informed choices.

Establish guidelines for assessing and mitigating potential biases in AI-driven systems, ensuring that they do not discriminate against specific demographic groups or perpetuate existing inequalities. This could involve regular audits and third-party assessments of AI algorithms used in the financial sector.

- *Right to explanation:* Grant customers the right to request and receive an explanation for AI-driven decisions that directly affect them, such as being denied a loan or being flagged for potential fraud. This would empower customers to challenge decisions they believe are unfair or incorrect.

Balancing Innovation and Regulation

Striking the right balance between encouraging innovation in AI and ensuring its responsible development and use is crucial for regulations to be accepted by all parties.

When drones became widely available, they were popular toys for those who could afford them. But complaints soon abounded of drones taking photos and videos of people and interfering with the airspace near airports, etc. In 2014, a woman in Seattle reported that a drone was hovering outside her apartment window, leading to fears of spying and invasion of privacy.[10] In 2015, a drone crash-landed on the White House lawn, no doubt causing the

Secret Service some serious alarm.[11] The drone's operator, a government employee, was off-duty and merely flying the drone for recreational purposes. Then, in December 2018, drone sightings near Gatwick Airport in the United Kingdom led to the cancellation of approximately 1,000 flights, significantly disrupting the travel plans of more than 140,000 passengers.[12]

These examples served as catalysts for new laws and regulations. The incident at the White House prompted the Federal Aviation Administration (FAA) to introduce mandatory registration for drone owners and stricter guidelines for their use. Incidents like the apartment drone in Seattle have been reported worldwide, leading to calls for increased regulation and the establishment of "no-fly zones" to protect individual privacy.

Many countries now have varying degrees of drone restrictions in place, trying to strike a balance between allowing the technology to flourish and not allowing it to trespass on existing rights. The regulations included provisions for how big the drones could be, that pilots had to maintain line of sight with the drones, and the drones had to be registered, licensed, and obey no-fly zones. Too much regulation can stifle incentives and opportunities for innovation by making the bar for development too expensive. The benefits of AI for societal well-being may never materialize. On the other hand, too little regulation can have harmful effects on consumers.

Another case of technology that needed regulation was the development of genetically modified organisms, commonly known as GMOs. For GMOs, some of the major regulations came in the form of labeling requirements, which would allow consumers to make choices about the food they consume. GMO developers are also required to monitor the health of consumers and the environment and report any adverse observations.

Critics of regulation complained that innovation would be curtailed. But proponents argued that a definitive set of rules establishing clear guidelines and legal framework would remove the uncertainties around drones and GMOs and give entrepreneurs and innovators more confidence in investing in these technologies. These regulations would also help with public trust and acceptance of the new technologies as the development of applications that address societal challenges would be encouraged.

These regulations often come about as a dialog between industry leaders and governments. One interesting proposal is to test out regulations by setting up regulatory sandboxes. For example, a regulatory sandbox is like a controlled testing area for new and innovative ideas in various industries, such as healthcare or technology. It allows companies and entrepreneurs to try out their products or services in a safe and supervised environment, where certain regulations are relaxed or adapted. The goal is to encourage innovation while ensuring that any potential risks or harms are carefully managed.

Examples of regulatory sandboxes at work abound in the financial technology industry. Several countries allow companies to pilot new services in a sandbox under close supervision before full market authorization. Technologies such as digital IDs and AI-powered decision-making can arrive safely to the public this way. The sandbox model has expanded beyond fintech as well. Sometimes, it has proven controversial, such as Canada's novel foods program, which aims to help entrepreneurs develop lab-grown and genetically modified foods.

Entrepreneurs have other guidelines they can follow; whether they strike a balance between stifling innovation and encouraging is not clear. The Cartagena Protocol on Biosafety is an international agreement, interestingly signed not in Cartagena but in Montreal, Quebec, that ensures the safe handling, transport, and use of living modified organisms (LMOs).[13,14] In the United States, the FDA's Biotechnology Consultation Program provides a voluntary consultation process for developers of genetically engineered plants, ensuring that the safety of food from these plants is thoroughly assessed before they enter the market.[15]

Economic and Social Impact

The march of technology is characterized by ever-changing demands on the workforce. Skills that are in great demand in one decade become historical artifacts in the next. The extensive infrastructure around horse-based transport that existed in the 19th century very quickly disappeared in the 20th century. Jobs such as farriers, stable hands, and coach drivers now exist mostly on farms and specialized sports. The widespread adoption of AI will change the skills that are required in many settings. The well-to-do, who have the resources to navigate and adapt to this changing skills landscape, will pull away even further from the have-nots. This is not a good outcome because increased economic inequality encourages increased social fragmentation.

Governments could provide incentives for companies to offer retraining programs for employees whose jobs are at risk due to AI-driven automation, ensuring they have the opportunity to develop new skills and transition to new roles.

What skills will be required for a new workforce that will need to collaborate with AI in their work? One truth is that the skills required three years from now might have been unheard of today. It is exciting for those who are well placed to navigate these waters but anxiety inducing for those trying to catch up.

Most people in the near future will not have to learn AI in the same way as today. They will not have to become proficient at coding, nor will they have to learn machine learning and statistics. But they will have to learn how to manipulate AI tools, acquire AI literacy, and, perhaps most importantly, understand the limitations of these tools. Specific AI-driven technologies play pivotal roles in augmenting operations. Whether it is robotic surgery within healthcare, adaptive learning systems within education, or autonomous vehicles within transportation, professionals must acquaint themselves with these technologies and possess a profound understanding of how to effectively collaborate with them.

Across various fields, the need for data analysis emerges as a common thread. Professionals must adeptly analyze substantial volumes of data to make informed decisions and drive improvements. Proficiency in AI skills, such as data analysis and machine learning, is instrumental in effectively processing and interpreting this wealth of information.

Ethical considerations loom large across all fields. Professionals working within these fields must grapple with ethical implications ranging from data privacy and fairness to transparency and accountability when implementing AI technologies. An awareness of AI's potential impact on their respective industries will be vital for the broader society.

Continuous skill development is imperative for professionals across all categories, driven by the need to remain abreast of technological advancements. This entails staying informed about AI tools, comprehending their practical applications, and leveraging them effectively in their work. Coursework offered by companies or sought out by individuals is already a practice in good use today.

While there are commonalities across industries, each category has its unique challenges related to AI adoption. For example, healthcare professionals must navigate the complex legal and regulatory landscape in healthcare, educators face the challenge of implementing AI in the classroom while ensuring inclusivity, and transportation workers must address public concerns about safety when using autonomous vehicles.

Let us examine a few of these industries.

Teachers and Educators

Many of the skills being taught in schools and universities are being automated. For example, generative AI tools such as ChatGPT can compose entire essays. This raises a question of whether educators should shift their focus to teaching prompt design for language models. Alternatively, should they prioritize equipping their students with the essential skills of constructing ideas and counterarguments needed for literary composition? The answer is somewhere in between, but probably closer to the latter. The trick will be to balance the disruption of Large Language Models with classical pedagogy.

How can educators incorporate AI into their curriculum? One potential avenue is in the adoption of personalized instruction, a method that may become more prevalent as intelligent tutoring systems continue to evolve. AI-driven tools can assist in tasks such as reporting, grading, and identifying learning gaps. Consequently, educators will need to possess the competence to teach students about AI concepts and applications, thereby fostering a generation of individuals who are well versed in the realm of AI.

Healthcare Professionals

The widespread adoption of AI-based decision aids is expected to become increasingly prevalent in healthcare.[16,17] For instance, the assessment of skin lesions, such as determining whether it is a form of skin cancer, can be entrusted to non-medical healthcare workers using AI models. An excellent illustration of this is diabetic retinopathy, where instead of visiting a doctor's office equipped with an expensive $100,000 camera to capture images of the retina, a healthcare worker can utilize a more affordable camera, such as a mobile phone camera right in their own village. As a result, more individuals can receive medical attention who may not have had access to it before.[18]

Healthcare professionals will need to become familiar with AI-driven technologies, such as robotic surgery, telemedicine, and virtual health assistants, as well as the ethical considerations that come with using AI in healthcare. And just like teachers and students, healthcare workers will have tools to analyze and query patient data. They will need the skill to interpret AI-generated recommendations.

Office Workers

Think of how computers have changed the office. While not completely paperless, as promised by tech moguls like Bill Gates in 1995, offices consume less paper each year, and carbon copies mainly live on in email headers. Many contracts are now signed online, archives are digital, and data analysis and reports rest on the back of Microsoft Excel. Communications are now over email, instant messaging, and video.

As AI develops, even more tasks will be automated. In the near future, we may witness AI systems taking on the responsibility of signing certain contracts. Customer service, too, might become largely automated, operating round-the-clock. As office workers witness the elimination or substantial reduction of many of their tasks due to AI, one might wonder whether this newfound efficiency will afford them more leisure time. However, historical evidence strongly suggests otherwise. Instead of maintaining productivity at its current level, employers are likely to view AI as an opportunity to increase production, assigning different tasks to employees to justify their continued employment. Consequently, lawyers may find themselves allocating more time to crafting arguments, while project managers could take on more projects and schedule more meetings, and personalized AI financial advisors may augment their human counterparts.

Manufacturers and Factory Workers

In manufacturing, AI will play a major role behind the scenes. The processes that govern machines will be dominated by AI systems. Just as automation and robotics have transformed manufacturing in developed countries, we should expect AI to spread this trend ever further. There will be a need for new skills for supervising and maintaining AI-powered machines. These might be mostly programming or configuring systems. The shift toward AI-driven manufacturing will redefine the nature of work for factory employees, requiring them to take on more strategic and creative tasks.

Traditional manufacturing jobs may give way to more automation. Quality control checks might be more efficiently done using AI-powered computer vision. This highlights the importance of investing in reskilling and retraining programs to help affected workers transition into new roles and industries.[19]

Transportation

Trucks travel very long distances, and their drivers need to plan their routes to minimize times while still hewing to several regulations about rest and speed. Companies such as Embark have tested coast-to-coast transport with an AI driving the truck. As the transportation and logistics industries move toward this type of automation, the infrastructure they have engendered will change as well. From charging stations to traffic management systems, we can expect to see the truck stop landscape, loading areas, etc., change significantly.

Truck drivers will need to become familiar with AI systems integrated into their trucks, allowing them to work alongside and supervise the autonomous driving technology. This requires a new skill set that includes technological literacy and the ability to effectively interact with AI-powered systems. Drivers will also need to develop expertise in monitoring and troubleshooting the AI systems to ensure smooth operations and intervene when necessary.

Farmers

Farmers should also be familiar with AI-driven technologies, such as precision agriculture, autonomous tractors, and crop monitoring systems.

Precision agriculture utilizes AI and sensor technologies to optimize farming practices at a granular level. Farmers will have to become familiar with AI-driven tools such as drones and ground-based sensors that collect data on soil conditions, moisture levels, crop health, and more.

AI-based crop monitoring systems use computer vision and machine learning algorithms to analyze images captured by drones or cameras mounted on agricultural equipment. They may also be able to alert farmers when crop diseases or nutrient deficiencies are identified by the AI systems. They should also be able to configure these devices and calibrate the AI, and then interpret the analysis results and take appropriate actions to optimize crop health and yield.

AI-powered autonomous tractors and machinery are transforming farming operations by automating tasks such as plowing, seeding, and harvesting. Farmers will have to acquire the skills to operate and maintain these autonomous systems as well as the ability to program and customize their operations based on specific field conditions and crop requirements.

Creative Professionals

If there's a group feeling threatened by AI, it is those in creative professions. Will animators still be indispensable for crafting Pixar and Disney films? Could we witness the creation of a robot capable of sculpting marble or will individuals simply print their own innovative 3D and 2D artworks? And what about music? It's not too far-fetched to imagine that the top of the music charts could soon feature one or two AI-generated compositions.

Tools like Midjourney, capable of generating images, might lead companies to outsource designs to AI or artists skilled in AI utilization. Similarly, architects may find themselves overseeing and refining AI-generated designs, and fashion designers may use AI for inspiration in creating new patterns, styles, and concepts.

Artists whose work has been used in training generative AIs may feel taken advantage of. In a sense, their own work has been used to put them out of business. Critics may argue that when art students attend school, they use similar materials for training without owing royalties to established artists.

There will certainly be a new wave of artists who are adept at using AI tools. Just as photography developed into an art form relatively recently, one can expect AI to generate novel arenas for artistic expression.

AI Governance

Governments have woken up to the necessity of regulating and putting guardrails around AI in order for it to develop fairly and equitably.[20] Policy making cannot keep up with the pace of technological development. As legislators were crafting laws, they were buffeted by the waves made by generative models like ChatGPT and Midjourney. They had to pivot quickly to address these as well. This realization, driven mainly by the success of generative models, doesn't mean that policies appear overnight. Governments are slow and deliberate in crafting effective legislation.

Regions worldwide are navigating the complex landscape of AI governance in a variety of ways. Europe leads the way with comprehensive regulations like the GDPR and the upcoming Artificial Intelligence Act.[21] The United States relies more on self-regulation, but some states have passed specific laws. China has taken a comprehensive approach focusing on data security, privacy, and transparency in algorithms. African countries are making progress with data protection laws and establishing task forces to shape AI governance.

The European Union's GDPR, which we encountered earlier, is one example that includes provisions related to AI, such as the right to explanation and data protection by design. Continuing as the leader of consumer protections, Europe is now crafting another set of laws collected into the Artificial Intelligence Act. It will categorize AI systems into specific risk classes ranging from minimal to systems with high risks and those that should be banned altogether. When it comes to AI systems making consequential decisions for people, especially high standards should be applied, particularly to the transparency of data a particular AI was trained for its decision-making that was used to train that system, and how the algorithms work to ultimately make decisions. There is hope that the protections and guidelines provided by this law will spread via the Brussels Effect to other parts of the world.

On the other hand, the United States currently has no legislation, at least at the federal level. Individual companies have taken to policing themselves, and have even told Congress that they would welcome regulation that covers all corporations and researchers equally. Individual states have some narrow laws targeting specific AI tools or uses, such as banning facial recognition by government surveillance. Illinois passed its Artificial Intelligence Video Interview Act, which requires interviewers to inform candidates if their interviews will be analyzed by AI.

China has been actively developing AI regulations with a focus on privacy, data security, and national security. These regulations exist alongside a national plan established in 2017 of becoming a global leader in AI by 2030. This plan emphasizes the importance of AI research and development and the integration of AI in all economic sectors.[22]

There is also the Cybersecurity Law and the Personal Information Protection Law, which are already in effect and intended to safeguard personal data. Regulations specific to facial recognition and biometric data usage are also in place, emphasizing the need for consent and classifying different levels of data security. For example, the Multi-Level Protection Scheme (MLPS) classifies different levels of data security, and the Cybersecurity Law imposes restrictions on collecting and using biometric data without consent. Guidelines and regulations are proposed to ensure transparency and accountability in AI algorithms, particularly in critical domains like finance and healthcare.

AI governance in Africa is still in the early stages, with countries at various levels of progress. Several African nations have enacted data protection and privacy laws that include provisions related to AI. For example, South Africa has the Protection of Personal Information Act, which governs personal data processing and encompasses AI systems. Some countries have established AI task forces and developed national strategies to guide AI development and address governance challenges.

In this chapter, we have looked at how AI may spill over our current established ethical and safety standards and which new challenges they bring up. Let us take a look at two areas that illustrate specific regulations that will be needed: military weapons and finance.

Military Weapons

When it comes to military weapons and the integration of artificial intelligence, establishing clear guidelines and limitations is crucial. Just like any societal endeavor, we need rules to ensure fairness and safety. In the arena of

AI regulation for military applications, several key considerations come into play.

First is the issue of autonomy and human control. It's important to strike a balance between AI capabilities and human decision-making. Regulations could require human oversight and control over AI systems used in military contexts. By doing so, critical decisions regarding the use of force can be made by humans rather than relying solely on AI algorithms. This helps prevent the potentially dangerous scenario of AI taking full control without any human intervention. It also helps keep the burden of human lives on human decision-makers, preventing what could be a blame-shifting scenario.

Ethics and accountability also come into play when we bring AI into the realm of military applications. The use of AI in this context raises complex ethical implications. Therefore, regulations need to address these concerns and emphasize principles, such as proportionality, distinction between combatants and noncombatants, and adherence to international humanitarian laws. AI systems should be monitored and trained to make sure the force they apply is not excessive and perhaps cause unnecessary harm or collateral damage. The goal is to ensure that AI systems are utilized in a responsible and accountable manner.

In addition to autonomy and ethics, another important aspect of AI regulation of military weapons is arms control and non-proliferation. We must guard against the unauthorized or inappropriate spread of AI-enabled military technologies. This entails establishing international norms and agreements to govern their development and use. Such measures can help prevent the misuse of AI technology by ensuring that it remains within the appropriate boundaries and under the control of responsible actors. Some have argued that this horse has already left the barn and irresponsible actors will be able to acquire AI technology. This makes it all the more important that international agreements include research and monitoring guidelines to discover misuse and international safety threats.

Finance

In finance, AI regulation aims to address challenges related to algorithmic trading, risk management, and consumer protection. We can identify three key aspects that such regulations may cover. Transparency and explainability promote understanding and clarity regarding AI algorithms. Fairness and bias mitigation ensure that AI systems do not discriminate or disproportionately impact certain groups. Data privacy and security measures protect customer data and guard against cyber threats.

Transparency and Explainability take center stage in AI regulation for the finance sector. Regulations may require financial institutions to be transparent about how they employ AI algorithms. This means ensuring that the decisions made by AI systems can be explained and understood. By promoting transparency, customers can gain insight into how AI is being used and have a clearer understanding of the reasoning behind financial decisions. There is a fine line between gaming the system and helping customers become better financial stewards.

Fairness and Bias are also critical considerations. AI regulations should aim to address biases and discrimination that may emerge in financial services. It is essential to ensure that AI algorithms do not disproportionately impact specific groups of people. By emphasizing fairness, these regulations work toward creating an equitable financial system, where decisions are made impartially and without discriminatory effects.

Data Privacy and Security hold significant importance in the realm of finance. Regulations will likely focus on protecting customer data and ensuring that financial institutions handle and utilize data responsibly. This means implementing robust measures to safeguard against data breaches and cyber threats. By prioritizing data privacy and security, regulations seek to safeguard sensitive financial information and provide customers with peace of mind regarding their personal data.

Final Thoughts

AI literacy is a step toward an understanding of the opportunities and perils that lie before us. In this way, the problem of AI development outpacing society's ability to absorb and regulate it might be mitigated. With better, more widespread understanding of what AI can do, we may be able to craft thoughtful policies.

It is not an exaggeration to say that the safety and well-being of humanity rests on how we collectively govern our use of technology. Whatever wisdom we have gained at the end of two world wars, economic recessions, and pandemics must guide us as we navigate these new terrains. A crucial part of this undertaking is to ensure a wide spectrum of voices is heard, from AI researchers who work with AI daily to the public at large.

This inclusive dialogue is essential for generating the momentum toward building a responsible and ethical framework for AI. It is dangerous to simply hope that someone will take care of it and look out for the rest of us. The task of AI governance is far too important to be left to chance or confined to technologists and lawmakers alone. This critical effort requires a collaborative, multidisciplinary approach that engages stakeholders from all sectors of society—government, industry, academia, civil society, and the general public.

References

1. Metz, C. "The Godfather of A.I." leaves Google and wans of danger ahead. *The New York Times* (2023). at <https://www.nytimes.com/2023/05/01/technology/ai-google-chatbot-engineer-quits-hinton.html>
2. A.E. Commission, USD Energy, US Government. *The Atomic Energy Commission and the History of Nuclear Energy: Official Histories from the Department of Energy - from the Discovery of Fission to Nuclear Power; Production of Early Nuclear Arsenal.* (Independently Published, 2017).
3. US Food and Drug Administration. Artificial Intelligence and Machine Learning (AI/ML)-Enabled Medical Devices. *FDA* (2022). at <https://www.fda.gov/medical-devices/software-medical-device-samd/artificial-intelligence-and-machine-learning-aiml-enabled-medical-devices>
4. Wiens, J., Price, W. N. & Sjoding, M. W. Diagnosing bias in data-driven algorithms for healthcare. *Nat Med* **26,** 25–26 (2020).
5. Wylie, C. *Mindf*ck: Inside Cambridge Analytica's Plot to Break the World.* (Profile Books, 2019).
6. Revell, T. How Facebook let a friend pass my data to Cambridge Analytica. *New Scientist* (2018). At <https://

www.newscientist.com/article/2166435-how-facebook-let-a-friend-pass-my-data-to-cambridge-analytica/>

7. Redman, T. C. & Waitman, R. M. Do you care about privacy as much as your customers do? *Harvard Business Review* (2020). At <https://hbr.org/2020/01/do-you-care-about-privacy-as-much-as-your-customers-do>
8. Helbing, D., Frey, B. S., Gigerenzer, G., Hafen, E., Hagner, M., Hofstetter, Y., Van Den Hoven, J., Zicari, R. V. & Zwitter, A. In *Towards Digital Enlightenment* (ed. Helbing, D.) 73–98 (Springer International Publishing, 2019). doi:10.1007/978-3-319-90869-4_7
9. Bradford, A. *The Brussels Effect: How the European Union Rules the World*. (Oxford University Press, 2020).
10. Woman calls police to report drone "spying" outside Seattle apartment. *KIRO 7 News Seattle* (2014). At <https://www.kiro7.com/news/woman-sees-drone-outside-apartment-window/81713615/>
11. Schmidt, M. S. & Shear, M. D. A drone, too small for radar to detect, rattles the White House. *The New York Times* (2015). At <https://www.nytimes.com/2015/01/27/us/white-house-drone.html>
12. Gatwick Airport: No drone found after flights disrupted. *BBC News* (2023). At <https://www.bbc.com/news/uk-england-sussex-65602504>
13. Bail, C., Falkner, R. & Marquard, H. *The Cartagena Protocol on Biosafety: reconciling trade in biotechnology with environment and development*. (Royal Institute of International Affairs, 2002).
14. Mackenzie, R., Burhenne-Guilmin, F., La Viña, A.G.M. & Werksman, J.D. in cooperation with Ascencio, A., Kinderlerer, J., Kummer, K. & Tapper, R. An *Explanatory Guide to the Cartagena Protocol on Biosafety*. (IUCN, Gland, Switzerland and Cambridge, UK, 2003).

15. Center for Food Safety and Applied Nutrition. Consultation Programs on Food from New Plant Varieties. *FDA* (2022). at <https://www.fda.gov/food/food-new-plant-varieties/consultation-programs-food-new-plant-varieties>

16. Mahajan, P. S. *Artificial Intelligence in Healthcare: AI, Machine Learning, and Deep and Intelligent Medicine Simplified for Everyone*. (MedMantra, LLC, 2021).

17. Bohr, A. & Memarzadeh, K. *Artificial Intelligence in Healthcare*. (Elsevier Science, 2020).

18. Trivedi, A., Desbiens, J., Gross, R., Gupta, S., Lavista Ferres, Dr. J. M. & Dodhia, R. Binary Mode Multinomial Deep Learning Model for more efficient automated diabetic retinopathy detection. In *Proceedings of the 33rd Conference on Neural Information Processing Systems 2019* (2019). At<https://www.microsoft.com/en-us/research/publication/binary-mode-multinomial-deep-learning-model-for-more-efficient-automated-diabetic-retinopathy-detection/>

19. Holzer, H. Understanding the impact of automation on workers, jobs, and wages. *Shifting Paradigms: Growth, Finance, Jobs, and Inequality in the Digital Economy*. (2022). At <https://www.brookings.edu/articles/understanding-the-impact-of-automation-on-workers-jobs-and-wages/>

20. Ghazal, I. An overview of global AI regulation and what's next. *Progressive Policy Institute* (2023). At <https://www.progressivepolicy.org/blogs/an-overview-and-of-global-ai-regulation-and-whats-next/>

21. AI Act: a step closer to the first rules on Artificial Intelligence | News | European Parliament. (2023). At <https://www.europarl.europa.eu/news/en/press-room/20230505IPR84904/ai-act-a-step-closer-to-the-first-rules-on-artificial-intelligence>

22. Webster, G., Creemers, R., Kania, E. & Triolo, P. Full translation: China's "New Generation Artificial Intelligence Development Plan" (2017). *DigiChina.* At <https://digichina.stanford.edu/work/full-translation-chinas-new-generation-artificial-intelligence-development-plan-2017/>

7

Getting the Best Out of Your AI Team

"Alone we can do so little; together we can do so much."
– Helen Keller

"Talent wins games, but teamwork and intelligence win championships."
– Michael Jordan

At the dawn of our current era of deep learning AI, IBM committed a blunder that has become a cautionary tale for big businesses' ambitions in artificial intelligence. In a 2011 episode of *Jeopardy*, one of the contestants was IBM's Watson supercomputer. They pit Watson against legendary contestant (and future host) Ken Jennings, whom it beat handily. Riding on that success, IBM executives were ready to promise even bigger things. AI would be an all-knowing assistant everywhere from office workers and farmers to oil prospectors and medical professionals.

They decided to start big and tackle cancer head-on, beginning with the renowned Memorial Sloan Kettering Cancer Center in Boston and MD Anderson Cancer Center in Houston. Drawing on the success of Watson on *Jeopardy*, which could quickly parse questions and provide accurate answers, they thought they could apply that same technology to cancer treatments and recommend personalized treatments to individuals.[1]

By feeding Watson all the medical literature they could and patient data, it would learn patterns and recognize the best treatments for given situations. However, cancer is a highly complex and dynamic disease, with numerous variables to consider, including patient characteristics, tumor characteristics, treatment guidelines, and evolving research. IBM Watson's ability to parse questions and provide quick answers demonstrated in its victory on *Jeopardy*, did not necessarily translate into accurate and reliable treatment recommendations for cancer patients. Physicians complained that it often made dangerous recommendations. The system struggled to keep up with the rapidly evolving advances in cancer research and lacked the necessary level of clinical expertise to make nuanced treatment decisions.

What led to the dismantling of this ambitious system? Was it the scientists who fed too many synthetic cases into Watson, the businesspeople who promised too much, the partners who didn't provide reliable data, or the doctors who didn't try hard enough? In hindsight, it is easy to point fingers. But given the moment, many people would have made similar decisions. Google and Amazon learned from these missteps. Subsequent assistants were much less ambitious in scope and Alexa and Google Assistant focused on smaller things. IBM, too, scaled down its NLP ambitions.[2,3]

But are we not at a similar place now with the emergence of ChatGPT? The excitement feels like the hype created after Watson's *Jeopardy* win. So, we should be cautious.

There have been other notable failures. Microsoft's Tay, discussed in Chapter 5, was an AI assistant that had to be shut down within three days. In the legal industry, AI has been hyped as just about to transform the industry for years. Being a notoriously conservative industry, that has not happened yet.

These failures are a subset of much larger project management and planning shortcomings. In this chapter, we will explore the principles that will greatly improve the chances of a successful AI project. It is not a primer on project management, even though many of the principles for AI projects are the same. We will focus specifically on the challenges that AI teams may face, particularly when engaged in AI for Good.

Roles in an AI Team

At the outset of any AI project, there are three major roles that may be spread across several individuals and sometimes, but rarely, concentrated within the same individual: the AI expert, the domain expert, and the project manager.

These three roles are necessary for the success of an AI project, but they are not sufficient. Above this team sits the executives, the people at the helm of the organization. Their material blessings and active support are crucial for the project to deliver real impact in the world rather than being just an interesting exercise. In addition to these, there exists a fourth role: the engineer. While the engineer role is also technical, it requires a set of skills distinct from those of the AI and domain experts.

A Three-Way Conversation

A successful AI project hinges on effective collaboration and communication among the persons embodying the roles of the project manager, domain expert, and AI expert. Let's examine what's expected of these three roles.

The Project Manager

To ensure the AI project stays on track and reaches its full potential, we need someone to coordinate everyone else. The project manager acts as the conductor, orchestrating the efforts of the domain and AI experts and ensuring that the project progresses smoothly. They are the linchpin that holds the team together, keeping everyone aligned with the project's goals and objectives.

They are logistics experts, planning wizards, and occasionally pulled in as counselors. They will create a roadmap with dates and deliverables and anticipate problems that might derail it. They are the glue that holds the team together, fostering effective communication among team members and other stakeholders. Budgets fall under their scope, as do ethics and communications.

In their role, the project manager actively engages with both the domain expert and the AI expert. By participating in conversations between the two experts, the project manager ensures that a shared understanding is established. They bridge any gaps in knowledge and expertise, fostering collaboration and synergy. This collaboration is crucial in setting realistic expectations for the project, as it allows the domain expert's insights to shape the AI expert's approach.

The project manager's interaction with the AI expert and the domain expert in an AI project can be likened to the manager of a restaurant working with the chef, restaurant critics, and customers.

In order to impress critics and bring in clientele, the restaurant wants to create a new dish. The manager, in the role of a project manager, is responsible for ensuring that the final dish meets the desired standards and aligns with the restaurant's vision. To do this, the restaurant manager must balance the chef's desire to expand culinary trends, the restaurant critics' and public's interests and tastes, as well as the restaurant's ability to procure and afford the new dish's ingredients.

Just as the project manager oversees the entire AI project, the restaurant manager oversees the culinary endeavor.

A project manager who understands the broad outlines of AI will be much more able to have productive conversations with the AI expert and the domain expert. Understanding the AI model will help the project manager communicate effectively with all stakeholders from managers to technical experts. They should always ask themselves three questions:

1. What are the model's input and output parameters?
2. What is the underlying mechanism of the AI model?
3. What are the limitations and potential risks of the AI model?

The Domain Expert

Every AI project targeted toward a real-world problem needs an authority in that field. They contain deep knowledge about that field, nuances, and demarcations that separate established knowledge from hypotheses not yet tested. They should know the business or research problem to be solved inside out. Their mission is to shape the AI project's objectives, ensuring it meets the unique needs of the project.

The domain expert, armed with their profound understanding of the industry or field, provides invaluable guidance to the AI expert. They bring practical knowledge and domain-specific context, ensuring that the AI model developed aligns with the project's objectives. For example, an expert in education may be able to say that a certain curriculum developed in Singapore would never work in the Middle East because of cultural differences. Or an expert in conservation could influence the type of AI model the AI expert builds, for example, by indicating that individuals within a herd of elephants need to be identified and tracked instead of the herd as a whole.

Collaborating closely with the AI expert, the domain expert shares insights on which parameters to trim, optimizing the model for the specific challenges it aims to address. By working together, the domain expert and the AI expert strike a balance between technical capabilities and practical applicability.

Continuing the restaurant analogy, think of the domain expert as a food critic or the restaurant's customers. The critic represents an external perspective, bringing in-depth knowledge of the food industry and providing valuable insights into the customers' preferences and trends. The critic provides feedback and evaluates the dish's quality, just as the domain expert offers insights and guidance to ensure the AI models align with the project's specific needs and challenges. Similarly, the customer represents the end user or beneficiary of the AI technology, providing valuable input and expectations that shape the project's direction.

The domain expert and the end user, the person who benefit from the model's final implementation, may not always be the same entities, but for this chapter we will assume they are the same.

Effective communication among all three parties is essential to produce the best outcome. Chefs willing to get feedback from critics and customers are more likely to succeed, and data scientists must be open to feedback from domain experts to create models that accurately reflect the problem area and deliver successful results.

However, this can be challenging as domain experts may have years or decades of experience, and communicating that in a short amount of time to a novice audience is difficult. The AI expert is likely a novice in the domain expert's field. AI experts must be willing to listen and learn from experts to ensure the project's success. AI experts and project managers may have to become minor experts in the domain they are working in, guided by their partners, the domain experts.

The interaction between the AI and domain experts may be fairly detailed. Let's say the AI expert has found a model that does pretty well but needs to make it smaller. With the domain expert guiding the AI expert on which parameters to trim and what types of output will be useful, the AI expert can help set realistic expectations.

The AI Expert

Residing at the heart of every AI project, there is a technical expert who possesses deep knowledge of artificial intelligence, machine learning, and programming languages. These languages will most likely be Python or

C++, though Java is also common. They will write the code that manifests the AI, and they will process the data so that the AI understands it. Then, they will evaluate the output of the AI models and report it so that the rest of the team understands how well the AI model is performing. Their mission is to ensure the AI project achieves its objectives with accuracy.

Think of the AI expert as the head chef. They possess deep culinary knowledge and expertise, similar to the technical expert in the AI project who has extensive knowledge of AI models and algorithms. The head chef has the skills, experience, and technical know-how of the appliances in the kitchen, and they have a great working knowledge of chemistry, how foods interact with each other, and how the application of heat changes matters. If they have access to good ingredients and equipment, their dishes will be tasty. And, it will be even more successful if they have good knowledge of what the restaurant's customers want and what critics expect.

Similarly, the technical expert develops and fine-tunes the AI models, optimizing them to achieve the desired results. And if they have good data and good communication with the domain expert, their models will be successful.

Setting Expectations About AI

The AI expert, with their technical acumen and expertise in artificial intelligence, explores various models and algorithms to identify the most promising one. Once a model is selected, they collaborate with the domain expert to fine-tune and tailor it to the project's requirements.

This collaborative effort ensures that the AI model is optimized and refined to deliver the desired outcomes. Additionally, the AI expert plays a crucial role in managing expectations by effectively communicating the capabilities and limitations of the AI system. By providing realistic insights into what the technology can achieve, they set the groundwork for reasonable expectations and prevent potential disappointments down the line.

As mentioned earlier, there are three critical questions that the project manager and the domain expert should consistently ask the AI expert.

1. What are the model's input and output parameters? With the AI model as the heart of the project, it is essential for the non-technical experts to know the data it requires and the output it will generate. This involves understanding the type and format of data that the model processes and the results generated. For example, the data's type could be images, audio, or text, and its format could be in flat files like Excel spreadsheets or any of the other numerous formats including specialist formats like DICOM, which is used extensively for imagining in the medical professions. By identifying these parameters, the project manager can facilitate effective communication between the AI and domain experts ensuring that the model is accurately designed and optimized to address the problem at hand.
2. What is the underlying mechanism of the AI model? Having a basic understanding of the AI model's underlying mechanism is valuable for the project

manager and domain expert. They don't need to grasp the intricate math or technical details, but having a general outline of the algorithm helps anticipate potential data and deployment challenges. The project manager will be able to troubleshoot issues more effectively and communicate efficiently with the team and other stakeholders.

3. What are the potential risks of the AI model? A responsible project manager should be aware of the model's limitations and potential biases. This involves recognizing situations where the model may not perform optimally. For instance, models built with data from a specific region might not generalize well to different regions. Additionally, in collaboration with the domain expert and AI expert, the project manager should assess the ethical implications of deploying the model. If the model has the potential to impact people's health or livelihoods, measures should be implemented to ensure it is not harmful.

The media has promised AI that can do wonders, albeit with a slightly skewed understanding of what's relevant. Movies also embed an idea of what may be possible. Even though we know when we're watching the movie that this is fiction, our memories are notoriously unreliable, and forgetting the source of that memory may cause us to remember it as real rather than fake.

For many AI enthusiasts who have not yet worked with the models and their data, the revelation that their complex problems may not be a straightforward task for an AI is akin to the realization that your favorite superhero is not invincible and can be defeated. So, an early conversation about what's possible with the available data is crucial for future success.

Effective communication is crucial at all stages of the project, but there are specific areas where AI experts may need to take the lead. The AI expert has to educate the domain expert on their specialty, particularly what AI is capable of and what they need in order to create a useful AI model.

Take a real-world example. A non-governmental organization has collected data on soil types, crop varieties, and plant genetics in several places around the world. This invaluable data, often available only to large agribusinesses, can be used to create food security and economic opportunities for smallholder farmers. The NGO can give these farmers advice about which crops to grow and when to grow them.

A potential client for them is Samuel, a farmer who speaks the local languages and cares for a large extended family. Like many other farmers, Samuel is always looking for ways to increase his profits. His ancestors grew maize and beans, but he's interested in exploring more profitable crops. However, access to information is limited by language barriers, lack of roads, and cell phone towers.

The NGO scientists read about AI and are intrigued. Perhaps AI can help them make even better recommendations. Maybe they can use AI to translate their recommendations into the local language and reach farmers like Samuel.

The AI expert can open their eyes to possibilities and help them set realistic expectations for their ambitions. The NGO scientists may have questions like these:

"Can the AI scour research articles for data on crops, fertilizers, etc.?"

"I've heard that AI can analyze my data and tell me what to plant and where to plant."

"I also want the farmer to send a photo and get an answer within a day as to what this pest is."

The AI expert's response to these will usually be centered around data availability, quality, and quantity. In our restaurant analogy, the public may love a particular seafood item, but the chef knows it is expensive and difficult to obtain. The restaurant manager knows it will cost too much to be profitable. By talking with restaurant critics, they may learn about new ingredient options being tried in other restaurants, and the chef can then incorporate those into their new dish.

Theoretically, with enough data, they can build models for most situations. The question the AI expert and their partners have to ask themselves is whether the cost of development, in terms of money and time, make sense. It's possible that the resulting model may also not be very good, and the AI expert may have a good sense of this before any work starts. After all, no one wants a bad model because any authority that turns out to be wrong will likely find themselves ignored.

The AI expert can also make their partners aware of other possibilities. Some typical answers are: "You don't need an AI model; traditional statistics are good enough"; "You can get free satellite images that will help you determine what the land is currently used for"; or "You can do this less expensively by using these open-source tools."

Case Study: Breast Cancer Example

In my team at Microsoft, many highly talented, versatile AI experts worked with researchers from medical institutions. One of them, Felipe Oviedo, partnered with researchers from the University of Washington and the Fred Hutch Cancer Center to improve the detection of breast cancer from MRI scans. It was an exciting journey of teamwork, communication, and scientific discovery.

The collaboration began with the two teams getting to know each other and understanding the strengths and expertise that each person brought to the table, not unlike two civilizations encountering each other after having heard rumors of the riches of their respective lands. This was during the pandemic when face-to-face meetings were unhealthy so the rapport was built online through video calls.

Introductions take the form of not just the title, but a description of what they do. Would you have any idea of what a "Medical Technical Strategist" is? They could be the person who presses the on/off switch on a monitoring machine, or they could be a person looking to improve recovery rates at a hospital.

At the heart of our approach is communication—a lot of it, because when two different minds meet, concepts are not going to be crystal clear. We acknowledge that our expertise lies in AI techniques not in diagnosing medical conditions from imaging or breast oncology. That is why we listen intently and carefully to understand the issues the oncologists are facing. Much later, we translate their concerns into statistical hypotheses and explore whether AI techniques could be a viable solution.

Once Felipe, the researcher from my lab, decided that the problem had potential for AI applications, he delved into logistics with his cancer research partners. This included understanding MRI scans and puzzling over how radiologists could make sense of the images and how the images should be fed to the AI model. The team considered whether data needed to be shared among institutions and what the privacy and security ramifications could be. The logistics and other details relevant to the project were documented in a project scoping document, which serves as a contract between all stakeholders, ensuring that all parties are working in step and toward a common goal.

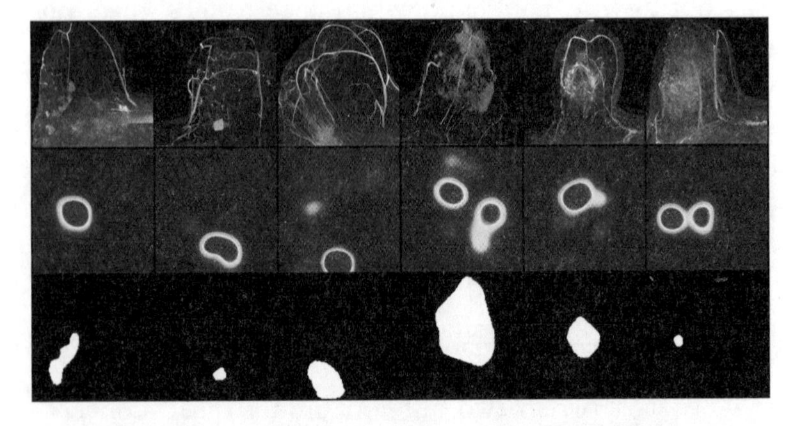

Figure 7.1 Example of breast cancer images. The top row shows MRI scans that a radiologist would examine. The middle row shows a heatmap of the areas the AI model considers to be anomalies. The bottom row shows radiologists' identification of anomalies. The model is consistent with the radiologist, and possibly more specific, in the sense that it highlights the most troublesome areas.

By establishing clear communication and addressing logistics upfront, we laid the foundation for a successful collaboration between the medical experts and our AI experts.

In this collaboration, the AI expert then dug into the data, meticulously analyzing the MRI scans and building deep learning models to understand the data's patterns and trends. The oncologists, project managers, and deep learning experts stayed in the loop by holding weekly touchpoints, where we shared our findings and discussed next steps. We found this frequency of communication, supplemented with specialized discussions throughout the week, is necessary for a true partnership across the team.

The touchpoints were not just about providing updates but also about actively seeking input and feedback from each other. It was important for the AI expert to understand the nuances of breast oncology and for the oncologists to grasp the capabilities and limitations of AI approaches. We made sure to create a comfortable environment for open communication, where both parties felt encouraged to share their insights, ideas, and concerns. This allowed for a dynamic exchange of knowledge and expertise sparking brainstorming sessions and discussions on how to proceed. We learned from each other's perspectives, and this dynamic exchange of ideas fueled the scientific process in action.

As the AI experts explored the data, they encountered some curious errors that their AI model was making. To address this, they took a selection of images that the AI struggled with and showed them to the oncologists. This led to insightful conversations, where drawing from their experience and knowledge, the oncologists suggested new approaches.

One such approach was to provide the model with pairs of breasts images so that information from one breast could inform diagnoses of the other breast. This novel idea emerged from the collaborative brainstorming sessions and highlighted the power of interdisciplinary collaboration. It was science in action, not following a perfectly planned-out roadmap, but rather choosing the right path based on collective expertise and insights.

Another challenge that emerged was the need for more examples of a particular type of tumor, which was not readily available in the existing data. The question arose: Could AI be used to create realistic fake images that would help the model learn what to look for? This sparked further discussions and exploration of new possibilities, leveraging the capabilities of AI to generate synthetic data for training.

Throughout the collaboration, constant communication was the key. The oncologists learned about the abilities and limitations of AI approaches while the AI expert gained a deeper understanding of what the oncologists were looking for in terms of cancer detection. It was a mutually beneficial relationship, where both parties learned from each other and pushed the boundaries of what was possible.

It was also crucial that the team consisted of highly motivated oncologists and talented AI experts who were dedicated to the cause. Their passion, expertise, and open-mindedness fostered a collaborative environment, where ideas were freely shared, challenges were collectively addressed, and innovative solutions were developed.

Project Scoping

The initial conversations among the AI expert, project manager, domain experts, and other key people are distilled into a project scoping document. A well-defined project scope helps set clear expectations, identifies potential challenges, and establishes a roadmap for achieving desired outcomes. How many times have you left a meeting that seemed to have everyone nodding and agreeing, and then different recollections muddy the results and endanger the project? Akira Kurosawa's movie *Rashomon* illustrates this effect, where different people have different understandings of what transpired during a shared event. A project scope, ideally written out in a document, is like a roadmap for achieving desired outcomes. It lays out assumptions, sets clear expectations, identifies resources and potential blockers, and provides a framework for decision-making.

Here is an example of a project's scope:

- **The problem statement:** The problem statement is the foundation of any AI project and should describe the specific issue that needs to be addressed. In one or more paragraphs, the domain expert should provide a description of the current state of affairs and the impact a solution would have. Even if the issue is high-level or vague, the problem statement itself should be specific and measurable. For example, an AI project about developing tools for finding survivors immediately after a mudflow or earthquake could be measured by the amount of time saved in finding people in specific situations.

- **The approach:** Describe what is needed to achieve the goal identified in the problem statement. In our example, this could be 1) identify areas where people are most likely to be buried using satellite or aircraft images, and 2) identify people using heat signatures. These statements help point the AI expert toward which AI approach to take. The modeling options should be listed here. Accompanying each of them should be a list of evaluation metrics. These metrics help the team determine whether or not their project is successful. The metrics should be relevant and meaningful. AI metrics often include accuracy, precision, recall, F1 score, AUC-ROC, and more.
- **Data sources:** The oft-used adage, garbage in garbage out holds true in AI as much as it does anywhere else. Before a project starts, diligent evaluation of the data is critical. Data that doesn't fit suitability criteria such as quantity, quality, and availability are a red flag. Potential biases in the datasets should also be highlighted for further investigation and mitigation. We go into these considerations more deeply later. In our example, data could be satellite images obtained from a commercial provider or aerial imagery obtained in data collection runs by aircraft.
- **Technical resources:** Technical resources are the tools and infrastructure needed to execute the project. This may include deploying drones to affected areas, accessing models in the cloud, considering where the model will be run, and where the data will be stored. Adequate technical resources ensure the project can be carried out efficiently and effectively.

- **Privacy, ethics, and responsible AI considerations:** Address any potential issues about privacy in the data upfront. Usually, PII should not be used unless there are specific licenses allowing it. Ethical considerations related to the use of data and technology should also be thoroughly evaluated ensuring that the project aligns with applicable laws and regulations. Proactively addressing these concerns in the project scope safeguards against potential risks and builds trust. Many organizations have placed a ban or severely restricted the use of face identification technology, for example.
- **Timeline:** Keep track by designating milestones and dates that can be reasonably expected.
- **Key stakeholders:** Names of the people doing the work and names of the organizations funding and using the work, with key contacts.

It is worth taking a lot of time in pondering and creating the project scope document. A common affliction for AI projects is when the end user is not quite clear on their desired outcome. This is often due to a limited understanding of AI capabilities. One way this affliction manifests is by engaging in an iterative process of exploration and refinement. Trial and error is how a lot of people problem solve from artists to computer programmers.

For example, consider an agricultural NGO as the end user. They have asked the AI expert and project manager for a landcover map of a region. The AI expert, trusting that the NGO will know how to make use of the output of their model, gives them a map of the region divided into regions of different land covers: cropland, urban areas, wooded areas, water bodies, etc. When the end user gets this, they realize that while it's pretty to look at and provides data and insights that were unobtainable before, there's not much they can do with it. They may find that they require a more detailed map that provides information on a per-acre basis. And once they see that updated output, they might realize that actionable output for them would be the ability to distinguish the crops that are grown or whether or not the land is arable.

This kind of iterative planning is fine when it is cheap. But it becomes unfeasible when teams are involved, with people to be paid and with expensive computer time. Shifting goalposts like this can be avoided by doing what-if scenarios before any work starts. The AI expert, being aware of the potential complexities involved, engages in a thoughtful exploration of the user's requirements. During the initial discussions, the AI expert and project manager can employ an iterative process to uncover more details about the desired outcome. They may ask questions such as:

1. What specific information are you looking to obtain from the landcover map?
2. How granular should the map be in terms of resolution or level of detail?
3. Are there any specific attributes or features of interest, such as distinguishing crop types or identifying arable land?

The Reality of Running AI: Cost, Connectivity, and Context

The AI team carefully considers the intended usage of the model, such as who the end users will be, the environmental conditions in which it will be deployed, and how much it will cost to run and maintain.

Will it be a model as large and complex as ChatGPT, which requires training at huge costs on multiple GPUs? Even though each individual query sets off a maelstrom of computation and communications across deep, buried cables, the cost of a query . . . 2 cents. ChatGPT is literally giving you its 2 cents. But multiply this by the millions of queries each day, and running these types of models becomes very expensive. No surprise then that only those with deep pockets or visions of large profits operate these models.

Or will the model that the AI team builds be designed to run on devices in the field? Just imagine the extension of AI technology to remote areas, where hospitals are scarce and even dispensing clinics are a rarity. Picture smallholder farmers working hard to cultivate crops and adapt to new diseases and techniques for improved yields. AI models exist that can help reduce the burden on medical professionals or improve the farmer's business. Still, they need to be compact enough to fit on a mobile phone and capable of smoothly running on the phone's low-powered CPUs without needing to be connected to the internet.

When the user is back in areas with mobile network connectivity, they can upload their findings from the phone and download new versions of the model. This approach enables the dissemination of valuable information and ensures that the model remains up to date.

There may be a future time when mobile networks that have wrapped most of the inhabited world in their electronic embrace will also be cheap. But right now, to a farmer in the field uploading a photo of a leaf with some blight on it might be expensive over mobile data networks. And mobile networks may not always be reliable.

Understanding the Role of Environmental Context in AI Deployment

In the pursuit of creating AI applications, project managers and developers mustn't overlook the importance of the environmental context in which their technology will be deployed. Ignoring this critical aspect could result in developers facing a runaway machine hell-bent on destruction. And nobody wants that.

To avoid such calamities, they must consider several factors, including the physical location of the AI system, the type of users interacting with it, and the impact of environmental factors on its performance.

Take as an example the problem of diagnosing diabetic retinopathy, a disease that affects hundreds of thousands of people globally. If you are lucky enough to be able to visit and afford an ophthalmologist, you'll find yourself seated in a chair with your chin on a chinrest, trying not to blink as a camera takes a detailed photo of your retina. The image can be diagnosed within a few minutes, and 1% of the people will get the news that they are in danger of diabetic retinopathy. Even if they receive this disquieting news, they will likely be able to take preventative measures.

They are fortunate to be living in a region where they could get an appointment and a quick diagnosis.

For the millions of people living in more rural areas, a similar diagnosis would involve a long and expensive journey to the doctor's office. Even if the doctor wanted to take himself to a central location to make the journey easier, there are only so many areas that can be served in this way. There is a shortage of medical professionals who can diagnose, let alone treat, diabetic retinopathy. As of 2021, there were about 200,000 doctors who could do this, and they need images taken by a camera.[4]

Therefore, health care workers need a solution that can work in the field. In South America, hospitals have been testing videos taken by ubiquitous mobile phones. How can a mobile phone camera possibly compare with the specialized equipment in a lab? This is the challenge many AI researchers have set themselves. The solution being tested in Mexico, Argentina, and Colombia requires a healthcare worker to take a video of the eye using a specialized app. There will be some techniques to be learned about holding the patient relatively still and moving the phone so that a useful video is captured. Then, an AI model scrutinizes each frame of the video to pick out the ones that capture the retina clearly. Another model further determines if there are signs of diabetic retinopathy. Via wi-fi or mobile networks, the app can then send the selected frames to physicians for a second opinion. This adaptation to the context requires an understanding of how the model will be used: in a rural setting, by healthcare workers that may have various levels of training, on patients with different lighting situations, and working within the confines of a mobile phone's capacity.

We find similar contextual challenges everywhere. When talking with African NGOs and participants at the major African AI conference, Indaba, a frequent theme is how AI models developed with a different population are transplanted to African situations. African organizations are now developing AI models with data that is grounded in Africa rather than relying on models built with data that may be irrelevant to Africa.[5,6]

The post–World War II Green Revolution in Europe and America improved food yields dramatically. Yet when it was exported to Africa, the results were not so great.[7–9] Many high-yielding seed varieties that had been developed successfully for temperate climates did not perform well in Africa's diverse ecological zones. Chemical fertilizers and pesticides, which were important for increasing productivity in the West, were often too costly for smallholder African farmers. The farmers who became dependent on seeds and fertilizers imported from the West fell into debt, exacerbating issues of poverty and inequality. The Green Revolution wisdom also directed that cash crops would be more beneficial for the farmers; maize or wheat could be exported. Traditional African crops that were better suited to local consumption and nutrition, such as millets, sorghum, and cassava, declined. With this came reduced resilience to climate and market shocks.

We see similar situations occurring with AI technology in Africa. Facial recognition technology developed in the West does very well with Caucasian skin tones and features. Like errors with darker skin tones in the West, the rates of misidentification among African populations were very high. Nevertheless, authorities such as the police jumped on this technology, promoting it as a tool to further their mission of law and order. Tests in countries like Zambia, South Africa, and Kenya take their cue from China, with the stated aim of reducing crime.[10, 11] There are strong concerns about this technology upending innocent people's lives by misidentifying them with technology developed using non-African faces.

Facial recognition technology has become a sensitive topic with civil rights and privacy advocates seeking to curb its use in the West. They fear that it may be used for social control and suppress dissent. In African contexts, where issues of governance, human rights, and social structures may differ from Western contexts, these problems are even more severe.

In a study conducted in Uganda, an AI-based mobile application was developed to provide health information and advice to expectant mothers. The application used natural language processing (NLP) and machine learning algorithms to provide personalized recommendations on prenatal care, nutrition, and maternal health. However, the study found that the AI-based application faced challenges in the local context including language barriers, low literacy rates among users, and discrepancies between Western medical knowledge and local cultural practices. The application struggled to provide relevant and culturally appropriate advice, and there were concerns about the accuracy and effectiveness of the AI-generated recommendations for expectant mothers in Uganda.

Similarly, healthcare settings pose a different set of environmental challenges that must be addressed. An AI system used in hospitals must adhere to stringent privacy and data security regulations to ensure patient confidentiality, like HIPAA compliance in the United States.

If the AI system makes use of a patient's private information, the developers will have to show how that information will be protected. Otherwise, they may join the ranks of hospitals and individuals who have had to pay millions of dollars in fines and losing their patients' trust.

Furthermore, the AI system must also integrate seamlessly with the existing IT infrastructure including electronic health records and other medical devices. This ensures that the system can exchange data without any hitches and operate as efficiently as possible. The story about IBM's failed venture with MD Anderson illustrates this. In the middle of the experiment, MD Anderson switched the system it used for electronic health records, rendering the IBM's oncology system defunct.

At a certain point, the marginal improvement in performance that comes from increasing the size and complexity of the model may not be worth the additional cost. For example, increasing the size of the model may result in only a small improvement in the quality of generated text but may also require a significant increase in computational resources and electricity usage.

Technology Resources

Another resource constraint is the availability of tech talent. Building a model is hard, and one that works reliably and accurately is even harder. Putting it into place—deploying it, as the tech jargon would say—requires an engineer who understands the software and can fix it when it breaks. AI models, like other types of engineering objects, work well most of the time. But the one sure thing is they will fail at some point. It's akin to predicting the weather; you can study the patterns and prepare as best as you can, but unexpected storms can still arise.

These causes could be mundane, like a traffic-predicting AI model encountering an unexpected slow-down due to a family of ducks crossing a road. Engineers may have anticipated many scenarios, but this specific event may have been deemed too unlikely to plan for.

A large company may have a team ready to jump in. Smaller organizations may be reliant on outsourced talent. Users in the field, say remote places such as Antarctica or the Amazon, may need to use something much more robust.

When the development of an AI model begins, have answers to these questions ready: Who will be responsible for maintaining the model? Do they have the necessary technical equipment and know-how? What downstream processes will be dependent on the model's success?

Data: Quantity and Quality, Annotations, Biases

Data is like a raw crystal, and the AI is how you shape and cut a diamond out of it. Most AI projects fail because of low-quality data.

Artificial intelligence projects rely heavily on data as the fuel that powers their algorithms and models. The type, quality, and quantity of data used in AI projects play a critical role in determining their success. In this section, we will explore the importance of these three key aspects of data for AI projects and highlight the need for data collection or data generation when there is insufficient quantity or quality of data.[12]

Type of Data: Format and Relevance

The type of data used in an AI project is crucial for its effectiveness. Different types of data, such as structured, unstructured, or multimedia data, require different approaches to processing and analysis. For instance, to predict famine in a region, one might make use of satellite images of river basins, historical flooding data, and even household incomes to determine the populations' resiliency. However, these datasets may exist in different formats, resolutions, or structures, posing challenges in data integration and alignment, and may require additional efforts for data preprocessing and standardization.

For another example, natural language processing (NLP) tasks require textual data while computer vision tasks rely on image or video data. The type of data used in an AI project should be carefully chosen to align with the project's objectives and desired outcomes. Using the right type of data ensures that the AI models are trained on relevant and meaningful information that leads to accurate and reliable results.

Quality of Data

The quality of data used in an AI project is paramount to its success. High-quality data is accurate, complete, consistent, and free from errors or biases. Poor-quality data can lead to unreliable and inaccurate results, compromising the performance of AI models. Data quality should be thoroughly evaluated during the data collection and preprocessing stages of an AI project. This includes identifying and addressing data inaccuracies, inconsistencies, or biases that may affect the performance of AI models. For example, imagine a dataset consisting of sound files used to train a model that detects the sounds of gunfire. If the dataset is mislabeled, for example, timestamps are annotated as having gunshots when they're nothing, or worse, a similar sound will make model training more difficult. It will make model training longer, less accurate, or both.

In my household, my wife and daughter are frequently frustrated by our smart speakers' inability to understand them. They are miraculously understood when they mimic deeper male voices. It's very likely that if the data used to train these smart speakers' models included more females, their experience would be improved.

Quantity of Data

The quantity of data used in an AI project also plays a crucial role in its success. Generally, larger quantities of data allow for more accurate and robust AI models. The critical factor here is not simply the amount of data. AI models learn best from diverse datasets, which have different examples to help it identify features that distinguish objects. A large dataset is more likely to consist of diverse examples. Trained with a diverse dataset, an AI model will perform better and generalize to new situations better.

Take a model designed to classify specific land utilization, such as distinguishing different types of vegetation across a vast expanse. Should this model be trained on satellite images of the East African savanna, it will do well in similar scenarios. When the AI expert uses that trained model on images of Argentinian pampas, she will notice a degradation in the model's performance. However, if the training data were more heterogeneous and included samples from a variety of regions, it would probably perform better in new scenarios. This is a major ongoing focus for AI research, finding methodologies for training models with minimal data or discovering the most effective mix of data to achieve optimal performance. The development of so-called zero-shot or few-shot models, which can detect unfamiliar samples without training on a specific dataset, was made possible by utilizing a broader array of exemplars.

Synthetic Data There are cases where data for an AI project runs scarce. In such situations, AI experts turn to a technique that vaguely feels like cheating. Called data generation, they create data rather than collect it. Using techniques to augment or synthesize data, existing samples can be used to create synthetic samples. Having thus called up the reserves and vanquished data scarcity, they can proceed with model development.

When encountering this technique, one may wonder why this "fake" data is not proscribed. Well, data synthesis has advantages that translate into better performance in the real world. The process of collecting data can often prove exorbitantly expensive, making the generation of new data from existing sources a cost-effective and valid alternative. Synthesis also proves valuable when existing data is unbalanced, which means that some important classes are inadequately represented. An example is the UK Biobank dataset, where 95% of the individuals are white and exhibit above average health conditions.

In situations where preserving privacy is important, creating synthetic data that emulates real data in broad strokes may be good enough for analysis. For example, there is a tradeoff between how much privacy you preserve with the synthetic data and the accuracy of the analyses. But sometimes, that tradeoff enables analyses that are otherwise not possible.[13] A good example here is a dataset released by Microsoft's AI for Good Lab that shows internet broadband coverage at the zipcode level across the United States. This dataset was crucial in determining actual coverage in the United States, uncovering estimates that significantly differed from those provided by the Federal Trade Commission, which relied on self-reports from internet service providers. To ensure the preservation of privacy, Microsoft researchers applied a data transformation technique that retained the statistical properties of the data while rendering it exceedingly challenging to identify individuals.[14]

In this example from Microsoft above, real data was transformed in a way that preserved the key measurements for the model. For many health research situations, the challenges of creating synthetic data can be quite complex. How do we know we are accurately representing the conditions being studied? Are there patterns in the synthetic data that researchers may be unaware of that AI will recognize and use erroneously? There are obviously important factors to consider when choosing synthetic data, so it's vital that all team members understand the possibilities and limitations and include model iterations and testing to ensure unanticipated patterns or biases are not being created or replicated.

Real-World Examples

In medical imaging, the MRI scans, CT scans, and PET scans are primarily conducted for diagnostic purposes, focusing on the patient rather than for machine learning–based research. Therefore, the quality that an AI expert would require is usually compromised. We often receive images that have annotations or marks in them that make them hard to process for use in AI models. For instance, there may be a ruler scale on the images to help doctors estimate how large a lesion or tumor is. The AI expert would have to find a way to remove that; otherwise, all that a neural network would learn is that the existence of a ruler means there is a tumor.

Another data flaw could be that the images are just not at a high enough resolution for the model to give accurate results. Or we may get a highly unbalanced dataset. When training a machine to identify lesions, it helps to have a good collection of images without any lesions to help teach it what a healthy image looks like.

Let's look at a couple of examples that highlight the challenges of data collection in the scientific world. They both involve videos, which to an AI is just a series of still images in a sequence.

The National Oceanic and Atmospheric Administration (NOAA) embarked on a project where they set up GoPro cameras at various underwater locations in the Puget Sound. The goal was to capture footage of marine life and count swimming and crawling creatures. However, we quickly realized that obtaining consistent data was no easy feat. The quality and background of the videos from different locations varied dramatically, presenting a significant hurdle for neural network modeling. It's like trying to solve a puzzle with mismatched pieces, making it difficult to draw meaningful conclusions from the data. In this case, as is described elsewhere, we did succeed in ameliorating the data problems.

Another project involved videos of babies. The quantity criterion seemed to be satisfied as there were a thousand videos. However, as the project progressed, we encountered unforeseen challenges. Some babies didn't move much due to their age or other factors, while the background in the videos was often inconsistent. As a result, the researchers were left with only a handful of usable videos. These unforeseeable complexities of working with seemingly simple data are unfortunately quite common. To a human who processes complex information with almost no conscious thought, the data seems rich—varied but consistent. But within the bounds of AI technology, we have to guess what will throw off a model.

These issues can sometimes be overcome. More data can be collected or require meticulous processing before it is ready for AI model training. The team, and often the leaders of the organization supervising them, have to balance incurring additional expenses with the potential benefits.

Modeling

Accuracy

Having more or less tamed the data beast, the AI expert is cautiously optimistic about coaxing promising results from it. However, their years of experience in academia, including graduate school, as well as their experience with real-world problems, instilled in them a prudent skepticism toward the output of statistical models. They have learned that while these models can provide precisely what is asked for, they may not necessarily deliver what the AI expert actually wanted. That is, these models can be overly compliant, providing results that strictly adhere to the input parameters without necessarily capturing the underlying context or addressing the nuances of the real world. Thus, their approach to analyzing and interpreting the outputs of statistical models is characterized by a thoughtful balance between utilizing their potential and acknowledging the limitations, informed by their extensive experience and critical mindset.

Accuracy: Balancing Precision and Sensitivity

When AI experts develop neural network models, they rely on scientific techniques honed over the last 300 years to determine if their creation is fit enough to see the light of day. The AI model is a statistical model and relies on statistical techniques, especially those developed in the 20th century. These techniques are used to compare the output of the AI model to outcomes in the real world, and if they match, the model is said to be reliable and effective.

Take species identification as an example. If you were to test a model that purports to identify the species of animal in images and gave the model an input of an image featuring an owl, the output would have to return owl as the species name. Deviations from the correct species, such as identifying it as a hawk or something unrelated like a bicycle, would be considered incorrect.

Some measures of accuracy do look at how semantically close the output is to the real answer, but many of the most common accuracy measures employ a binary decision.

When evaluating the accuracy of the model, there are two perspectives on what that may entail. Take as an example a neural network model that is 90% accurate at identifying owls. One interpretation of that statement could be that when the model is presented with 200 photos split equally between owls and non-owls, the model correctly identifies 90 owls. This accuracy measure is called *Recall or Sensitivity*. A model with perfect recall would have identified the 100 owl images correctly. But if we were to rely on just recall, that wouldn't be very informative. For example, an idiotic model that says everything is an owl would still have perfect recall! Therefore, there is a complementary measure of recall that considers the non-owl images. This measure, called *Precision*, counts how many of the images classified as owls by the model were actually owl images. So, if the model said 80 images were owls but in reality only 50 of them were owls, then 30 out of the 80 were false identifications. In this case, we have a recall of 50% (because out of the 100 owl images, 50 were identified as owls) and a precision of 50/80 = 62.5% (because of the 80 images identified as owls, only 50 were actually owls).

What do those values actually mean? Is a precision of roughly 63% good or bad? That is up to the judgment of the AI and domain experts and the demands of downstream tasks. Some fields, such as physics, require extremely high measures of precision and recall. In other fields, such as wildlife conservation, a high recall coupled with a lower precision may be acceptable. Here, it's more important to cover as much ground or capture images of as many animals as possible, even if some of them turn out not to be relevant.

While recall focuses on the model's accuracy in correctly identifying the target species, precision highlights the model's ability to avoid misclassifications particularly when dealing with different species. Both measures are valuable in assessing the model's usefulness. You may encounter other terms for accuracy such as sensitivity and specificity. They serve the same purpose, evaluating a model two ways, both with how good it is at identifying the objects of interest and how bad it is at misidentifying them.

These measures are applied to testing classification models that judge countable things. When you ask a scientist how good their model is, expect them to give you two numbers, sensitivity and precision. If you see only a measure called *accuracy* for the model, you should ask for the model's precision and sensitivity values as well.

Human-in-the-loop

AI experts, like others working on complex mathematical analyses, strive to create models that are accurate and will not break when they encounter niche cases. This takes a lot of work.

Remember the feeling of awe and delight when you first experienced your phone reliably transcribe your speech into text or when a search engine guessed what you really meant to type despite your less than perfect spelling? Although routine usage of these miracles may have dulled their initial impact, taking a moment to reflect inwardly will reignite a sense of amazement when you realize that these technological feats were almost inconceivable merely a decade ago.

These remarkable advancements are made possible by the power of artificial intelligence. The technology seems elegant and powerful on the surface. But behind the scenes are thousands of tech workers, engineers, product managers, and scientists tweaking and refining these models, getting rewarded by watching as accuracy inching up from 95% to 95.5% then to 96%. With a planetful of people testing their creations, it is inevitable that bugs will be found and will require fixing. With some of these users actively seeking weaknesses that might infiltrate their code, some are bound to succeed and cause much deeper security concerns.

These systems are not set-and-forget; they require continuous attention and adaptation to keep up with the changing world. Take Google maps, for example. As the world changes, its underlying data is changing. Maybe someone will report that there is now a motorway where there only used to be a dirt road. Or that the translation of the term 躺 in Chinese (to lie down) is rendered incorrectly in Japanese as 噓 (to tell a falsehood).[15] These armies of techies will be tasked with a tsunami of fixes, improvements, and testing alternatives. If this sounds a bit Wizard-of Oz-ish, it isn't really. We have a technical term for it called human-in-the-loop. This puts a human, often a domain expert, as the overseer of the AI, carefully supervising its output before letting users set eyes on it. The role of these human experts is crucial in ensuring the accuracy and ethical behavior of the entire system.

In a medical diagnosis scenario, for example, it is unlikely that doctors will rely solely on the AI model's diagnosis and will examine the eye images themselves. What they might do is prioritize patients based on order of severity as determined by the AI model. The involvement of human judgment is driven by concerns surrounding accountability, which were discussed in Chapter 6. Recognizing the potential impact of medical decisions on patients' well-being, doctors understand the need to personally review and validate the AI model's results. By doing so, they assume responsibility for the accuracy and reliability of the diagnosis.[16,17]

Consequences for Wrong Outcomes Are More Severe

Product managers and machine learning scientists soon learn they have to strike a balance between precision and sensitivity. Consider a model that identifies cancers from MRI scans. The physician supervising such a model would want it to be conservative, in the sense that it should not risk missing a cancer signal in any image. Therefore, its sensitivity is set to a very high level. If there is even a small chance that an image shows cancer, it will put it in the "has cancer" pile.

Because of the model's high sensitivity, a lot of the images that the physician reviews will not have cancer. This doesn't seem to be a problem because the physician would want to err on the side of caution. However, there is a cascading effect. With so many false positives and low precision, trust in the model erodes. People will be sent in for treatments and more tests that put strain on resources.

Eventually an administrator will say these resources are misallocated and create inefficiencies in the health system. Inefficiencies lead to higher costs and possibly lower quality outcomes for some people who actually do have cancer.

So, the AI expert turns the dial and makes the model less risk averse. That is, sensitivity is dialed down and precision is dialed up. The threshold is higher for the model to decide to put an image in the "has cancer" pile. The model will have fewer images in that pile, but the doctors will be more likely to believe the model. They'll find that an image almost certainly shows cancer if the model flagged it. As you've no doubt realized, a lot of people who might have had cancer will not make it through the model's diagnosis. And unless someone else scans these images, they might be too late in finding out that they have cancer.

The dilemma for the AI, domain, and project manager experts is to find a balance where the AI model is saving time to diagnosis and also not letting cancer slip through undetected.

A similar model using computer vision techniques might be used for identifying people buried under rubble after an earthquake. Here, a model's sensitivity must be dialed up very high. False alarms are acceptable so long as all possible survivors are found.

However, when it comes to convicting someone and adhering to the principle of "innocent until proven guilty," a model providing guilt recommendations must exhibit great precision, meaning it does not often falsely announce guilt. Consequently, the rate of false negatives, which involves the risk of letting an actual criminal go free, increases. This tradeoff raises fundamental questions that society must address in accordance with its ethos.

Professions that have a significant impact on our well-being and finances, such as medicine and law, tend to adopt a conservative approach, and rightfully so. Interestingly, the banking sector appears to be comparatively more laissez-faire. However, this contrast likely stems from a careful analysis of risk and cost. For instance, banks may implement policies that allow customers to access their accounts over the phone by verifying their voice. This approach offers convenience and is likely to attract numerous new customers. However, AI can easily imitate your voice, posing a potential security vulnerability.

The substantial profits generated by the banking industry often lead to an acceptance of some casualties. This tradeoff between convenience and security is driven by the economic considerations of the industry. While it may result in a few individuals falling victim to fraudulent activities, the overall financial gains for the banks outweigh the associated costs. Individuals who have become victims may have differing opinions from the banks about whether these are acceptable losses. This is where regulations as those discussed in Chapter 6 may be considered.

How could an AI be used in the legal profession? When a merger is in the works, it could determine which documents are risky. Or when sales agreements are being negotiated, it could generate clauses based on what has passed muster before with similar customers.

In the event that an AI erroneously assesses the risk of a document or incorrectly specifies a clause, the consequences could be dire, potentially resulting in a hapless attorney's sacking and significant financial losses for the corporation. While the corporation itself will not go to prison, the financial impact of such an error would be substantial. Consequently, it is critical that the AI is viewed as an aide with a human-in-the-loop paralegal, technician, doctor, or junior attorney serving as a vigilant overseer carefully scrutinizing the AI's work to ensure its accuracy and reliability.

Of course, there are still limitations to what AI can accomplish in the legal profession. For instance, there are many legal tasks that require the judgment and discretion of a human expert, such as evaluating the strength of a case or negotiating complex settlements. In these instances, AI can still be useful as a tool to assist human lawyers, but it cannot fully replace them.

Final Thoughts

While the fundamental nature of AI teams will largely remain unchanged, one particular role is likely to undergo significant transformations—that of the AI expert. As humanity ventures further on its AI journey, we will see the technology change, and we consider two of these potential changes in the final chapter. Project managers and domain experts will continue to play crucial roles in their respective fields, but AI experts will need to acquire new skills to adapt to the evolving landscape.

As AI progresses, the knowledge and expertise possessed by domain experts will increasingly overlap with those traditionally associated with AI experts. This convergence is expected to lead to a shift in responsibilities, where domain experts themselves will assume the role of AI experts at some point in the future.

The transition from the wild days of AI to a more regulated and constrained environment may introduce new challenges, but it also presents opportunities for the growth and development of AI professionals. By embracing these changes and adapting to the shifting demands of the AI journey, both domain experts and AI experts can contribute to the responsible and impactful advancement of artificial intelligence.

References

1. Greenstein, S., Martin, M. & Agaian, S. IBM Watson at MD Anderson Cancer Center. *Harward Bus. Shool* 9–621 (2021).
2. Lohr, S. Whatever happened to IBM's Watson? *The New York Times* (2021).
3. Best, J. IBM Watson: The inside story of how the Jeopardy-winning supercomputer was born, and what it wants to do next. *TechRepublic.* https://www.techrepublic.com/article/ibm-watson-the-inside-story-of-how-the-jeopardy-winning-supercomputer-was-born-and-what-it-wants-to-do-next/ (2013).
4. Trivedi, A., Desbiens, J., Gross, R., Gupta, S., Lavista Ferres, Dr. J. M. & Dodhia, R. Binary Mode Multinomial Deep Learning Model for more efficient Automated Diabetic Retinopathy Detection. In *Proceedings of the 33rd Conference on Neural Information Processing Systems 2019* (2019).
5. Wairegi, A., Omino, M. & Rutenberg, I. AI in Africa: Framing AI through an African lens. *Commun. Technol. Dév.* (2021) doi:10.4000/ctd.4775.
6. Ade-Ibijola, A. & Okonkwo, C. Artificial intelligence in Africa: Emerging challenges. In *Responsible AI in Africa: Challenges and Opportunities* (eds. Eke, D. O., Wakunuma,

K. & Akintoye, S.) 101–117 (Springer International Publishing, 2023). doi:10.1007/978-3-031-08215-3_5.
7. Dawson, N., Martin, A. & Sikor, T. Green revolution in sub-Saharan Africa: Implications of imposed innovations for the wellbeing of rural smallholders. *World Dev.* **78**, 204–218 (2016).
8. Daño, E. C. *Unmasking the New Green Revolution in Africa: Motives, Players and Dynamics.* (Third World Network, 2007).
9. Francesca de Gasparis and Fletcher Harper, opinion contributors. America's green revolution is failing African farmers. *The Hill.* https://thehill.com/opinion/energy-environment/3509220-americas-green-revolution-is-failing-african-farmers/ (2022).
10. Hawkins, A. Beijing's big brother tech needs African faces. *Foreign Policy.* https://foreignpolicy.com/2018/07/24/beijings-big-brother-tech-needs-african-faces/ (2018).
11. ENACTAfrica.org. Who's watching who? Biometric surveillance in Kenya and South Africa. *ENACT Africa.* https://enactafrica.org/research/research-papers/whos-watching-who-biometric-surveillance-in-kenya-and-south-africa (2020).
12. Kshirsagar, M., Robinson, C., Yang, S., Gholami, S., Klyuzhin, I., Mukherjee, S., Nasir, M., Ortiz, A., Oviedo, F., Tanner, D., Trivedi, A., Xu, Y., Zhong, M., Dilkina, B., Dodhia, R. & Lavista Ferres, J. M. Becoming good at AI for Good. In *Proceedings of the 2021 AAAI/ACM Conference on AI, Ethics, and Society* 664–673 (Association for Computing Machinery, 2021). doi:10.1145/3461702.3462599.
13. Kapelke, C. Using differential privacy to harness big data and preserve privacy. *Brookings Institution.* https://www.brookings.edu/articles/using-differential-privacy-to-harness-big-data-and-preserve-privacy/ (2020).

14. Pereira, M., Kim, A., Allen, J., White, K., Ferres, J. L. & Dodhia, R. U.S. Broadband Coverage Data Set: A Differentially Private Data Release. Preprint at https://doi.org/10.48550/arXiv.2103.14035 (2021).

15. impurekitkat. Google Translate caught red handed: Chinese is translated to English, then to Japanese?? *r/LearnJapanese* www.reddit.com/r/LearnJapanese/comments/13fvyxt/google_translate_caught_red_handed_chinese_is/ (2023).

16. Grote, T. & Berens, P. On the ethics of algorithmic decision-making in healthcare. *J. Med. Ethics* **46**, 205–211 (2020).

17. Mahajan, P. S. *Artificial Intelligence in Healthcare: AI, Machine Learning, and Deep and Intelligent Medicine Simplified for Everyone*. (MedMantra, LLC, 2021).

8

The Future

"I always avoid prophesying beforehand because it is much better to prophesy after the event has already taken place."
– Winston S. Churchill

"The old order changeth, yielding place to new."
– Alfred Lord Tennyson

THE FUTURE OF artificial intelligence in the hands of its human masters is large and bright, and admittedly accompanied by more than a twinge of trepidation. Similar emotions must have accompanied earlier breakthroughs in human society. When the telephone and the radio were invented, prognosticators of the time could promise widespread faster communication and easier access to information, but the deeper societal transformation they would bring about was much harder to anticipate. Global convulsions occurred as they enabled wars to be coordinated on a worldwide scale. At the same time, these technologies fostered a sense of unity and shared experience. For the first time, entire nations could tune in to the same broadcasts and share common narratives.

Now, with artificial intelligence, we are at a similar pivot in history, but something is different. Machines have reached a stage where they are no longer confined to replicating human motor functions. They are reaching deeper into human essence and replicating how we think and imagine. Though still in its infancy, the swift progress of artificial intelligence we've seen so far foretells the emergence of systems capable of creativity, logic, and problem-solving skills that were once considered solely human attributes. In this chapter, we begin to uncover what this means for humanity.

New Technologies

There are new technologies coming up that will boost AI beyond what we can imagine. Our situation in history can be compared to when steam and electricity were automating the world or when long-distance communication was still seen as a magical novelty. Two of these upcoming technologies warrant a closer look because of their potential to bring about a paradigm shift in AI and unlock unprecedented achievements. They are quantum computing and DNA storage.

Quantum Computing

AI flourished because chip makers figured out how to make faster, more powerful processors and denser, more reliable storage. For decades, they came up with better methods for squeezing more circuitry onto tiny wafers of silicon. Today's processors, able to work on multiple tasks simultaneously, can perform trillions of operations per second. If you had a cent for each operation, you'd be on *Forbes* list of the world's richest people. Back in the 1960s, higher-end mainframe computers achieved 750,000 operations per second. Using the same analogy, even accounting for inflation, you'd have about $75,000.

For six decades, Moore's Law, which predicted the exponential growth of transistor density on integrated circuits, served as a guiding principle for chips. But the law, which is not a law in the sense of a physical law, is now pushing against the boundaries of physics. The chips are now so small and the circuitry so thin that pushing electrons around reliably is becoming harder. Chip makers are now faced with the challenge of finding alternative ways to sustain the exponential growth that has become synonymous with technological progress.

Several technologies are already in the research phase, and this century's innovations will be even more fantastic than what we've seen so far. The first one represents a paradigm shift in computational power. The very nature of how computation occurs and the substrate upon which it will be realized might appear as weird to us as the internet would have appeared to renaissance people. It is called quantum computing. As the name suggests, it uses quantum effects at the sub-atomic level to achieve computation speeds that are astounding.[1]

In classical computing, the fundamental unit of computation is a bit, which can either be 0 or 1. However, in the realm of quantum computing, the basic computational unit is a qubit, which bizarrely can represent both 0 and 1, simultaneously. This seemingly contradictory situation arises from the principles of quantum physics, specifically the concept of superposition. This principle has puzzled scientists since Schrödinger described his thought experiment involving a cat inside a box with a potentially lethal gas that could be released based on a random event. To an observer outside the box, the cat appears to be both alive and dead at the same time—simultaneously one and zero.

On chips manufactured today, transistors are used to represent binary information. While it is possible that future technology may develop qubits on silicon chips, current research is exploring alternative methods for working with qubits.[2] Some of these approaches include trapping ions within an electric field and manipulating them using lasers or employing quantum dots, which are extremely tiny semiconductors that can be tuned to emit light at different wavelengths. As we continue to push the boundaries of computing technology, these innovative techniques may pave the way for even more remarkable advancements in artificial intelligence and other fields.

Although potentially years or even decades away from becoming a reality, the quantum computing revolution is worth keeping a close eye on. This is because the implications of this radical shift in computing are already evident. Because a quantum computer can work on problems in parallel instead of sequentially like a classical computer, it will be able to punch through challenges that appear intractable to classical computers.

Like encryption, which underpins secure financial transactions between banks and individuals, data sent from one computer to another across the globe is encoded into a format that would appear as gibberish if one were to display it on a screen. This complex encoding process, a descendant of codes and ciphers developed since ancient times, ensures that sensitive information remains secure and inaccessible to unauthorized parties. The banking system survives because these encryptions are almost impossible for anyone to break without access to the keys that can decode them. The strength of encryption relies on the difficulty of factoring very large numbers—a task that takes classical computers an incredibly long time to accomplish. It is related to a hacker trying to guess your password. For example, a password that has 8 characters might take less than an hour to crack using a GPU, whereas a 16-character password could take several billion years.[3] These estimates are based on a brute force approach that involves testing every combination of characters.

Quantum computers with the correct configuration of qubits could potentially excel at cracking codes like these within a time frame that a malcontent would find reasonable. You don't have to squirrel away your savings under a mattress just yet. This technology is several decades away, and encryption schemes that will be hard for even quantum computers will be devised. However, its contrast with classical computing shows its power.

Combining the power of neural networks with the power of quantum computing appears to be a recipe for unimaginable intoxication. We could witness AIs being trained with parameters that could potentially reach quintillions and more in number—a stark contrast to the mere trillions we see in the larger models today. That is if we continue with the current architectures of neural network models.

This prospect is not just mind-boggling; it is truly vertiginous. If we are already marveling at sparks of human-level intelligence in the elementary AI models of today, one can barely fathom the potential aptitudes of AIs built on an entirely alien scale and architecture.

DNA Storage

AI as we see it today became possible because of our ability to gather and process vast amounts of data. Data is like the lifeblood of AI. The more data an AI system can learn from, the smarter and more accurate it becomes. With large and diverse datasets, AI algorithms can recognize patterns, understand language, identify objects, and even predict future outcomes. This ability to learn from data is what allows AI to perform tasks that were once thought to be possible only for humans.

AI powered by quantum computation will be stillborn if it is not fed with corresponding increases in training data. Without sufficient and diverse training data, quantum AI will not be able to reach its full potential. This data will also need to be stored somewhere that can be easily accessible, and in vast quantities.

There are constant breakthroughs in developing denser and more reliable storage solutions to accommodate the massive amounts of data required for AI applications. In the early days of computing, vacuum tubes were used for data storage, which were bulky, power-hungry, and limited in capacity. Over time, advancements led to the development of various storage technologies, such as magnetic tapes, hard disk drives (HDDs), and solid-state drives (SSDs). Flash memory, as found in USB drives and, quite possibly, the SSDs in your laptop, revolutionized data storage with its fast reading and writing speed, reliability, and high storage capacity, which can be measured in terabytes.

Technology for memory marches forward, but one stands out in its exoticness and potential: using DNA as a storage medium. This concept draws inspiration from the remarkable way the code to create living organisms is held in a special type of molecule.

Living things are created of molecules that are made from atoms found plentifully on Earth. These molecules come together to form us—our skin and bones, brain and blood, everything that makes us who we are. Their ability to organize themselves into these complex structures is based on autonomous processes directed by our genes.

Genes, in turn, are created by our DNA, which is present in strands within each of our cells. The strands are incredibly long and hold vast amounts of information. One would imagine they'd have to since they hold the information to create a whole complex living being.

Someone had the bright idea that DNA could be used to store other information as well. For example, a feature film could be encoded into the basic building blocks of DNA, and then it could be retrieved the same way genes and proteins are retrieved from DNA. Sounds far-fetched? It has already been done. A team at Harvard University led by George Church published a groundbreaking paper in 2012 titled "Next-Generation Digital Information Storage in DNA" in the journal *Science*.[4] A few years later, a short movie was encoded into DNA, retrieved, and replayed.[5]

How is DNA moved from place to place? Right now, it's in little glass vials. A gram of liquid in that vial measuring about 1 cubic cm in volume (roughly the size of a standard sugar cube) would be capable of holding over 200 petabytes of information.[6] That is like having 200 very large libraries filled with books from floor to ceiling.

However, like quantum computing, there are enormous technological challenges yet to be overcome. The current state of DNA storage technology suggests that it may be a great medium for archival purposes, but for uses that require rapid reading or writing, it is far from suitable. But this is now, and one can hope that the merger of molecular biology and computer science will yield breakthroughs that make the speed of retrieval and storage in DNA faster than current methods.

AI Teams in the Near Future

While we await society's reaction to these new frontiers, we have shorter term adaptations to engage in. The ability to embrace change and learn new technologies will be vital for professionals. AI systems and tools will continue to evolve, requiring individuals to proactively upskill and reskill to remain relevant in the job market.

In Chapter 7, "Getting the Best Out of Your AI Team," I claimed that, unlike AI experts, project managers and domain experts will not have to adapt their skills as drastically often. However, they must still adapt to new AI technologies to remain competitive. These tools will automate many mundane administrative tasks, such as scheduling meetings, task assignments, tracking project progress, and monitoring team performance. This automation will free up significant time for project managers, allowing them to focus on more strategic aspects of their role.

The shift in project management skills might see them taking up more of a people-centric role. As AI takes over procedural tasks, project managers can focus more on fostering a culture of teamwork, empathy, and continuous learning within their teams. They will also need to manage change effectively, guiding their teams through the technological transitions while minimizing disruption to project timelines and outcomes. Ethical considerations related to AI use will become a core part of their responsibilities. They will be instrumental in ensuring that AI is implemented responsibly and transparently in their projects. This includes considerations about data privacy, AI fairness, and potential impacts of AI decision-making.

Domain experts will also find that AI has taken over some of the more mundane part of their jobs, which will include finding insights in data and generating reports and graphs. It is important for domain experts to remember their unique value. While AI can provide data analysis, the interpretation of the results and application to real-world contexts requires human judgment and expertise. The partnership between domain experts and AI should be a symbiotic one. Domain experts should see AI as a tool or an aid rather than an infallible authority.

Think of some tasks that could be automated by a reasonable intelligence able to handle vast details. For example, scheduling and logistics for complex organizations, such as NGOs' planning for volunteers responding to disasters. Or the air transportation industry juggling factors, such as crew availability, aircraft maintenance schedules, weather conditions, and airport slots. An AI system could automate this process analyzing all these variables in real time to optimize schedules and ensure smooth operations. If a crew member falls ill or a plane needs unexpected maintenance, rather than rely on hard-coded rules or decision-making rules of thumb the AI could quickly adjust the schedule and minimize disruptions. Human schedulers and planners already use data-based decision aids to help them coordinate this vast machine, but they will find these tools changing as AI permeates them.

In medicine, we have already seen how radiology and pathology will help physicians make earlier and better diagnoses. But surgeons, too, could benefit in the operating room. A robotic surgeon could assist in suturing wounds, making precise incisions, or even performing entire procedures under the supervision of a human surgeon. In post-operative care, AI could assist nurses by monitoring patients' vital signs and alerting healthcare professionals to any changes that might indicate complications. This could help to catch issues early, before they become serious, improving patient outcomes. The people these AIs will assist, the nurses, surgeons, etc., will have to become very familiar with their AI aides, and their experience will be invaluable in honing the AIs' abilities.

AI-Specific Jobs

When computers became integrated into society, new industries were born around the new technology. Adepts likened themselves to wizards and gurus because, frankly, they seemed to be able to do things with keyboards and green-glowing screens that seemed magical.

Then, the internet spawned a host of new job titles, such as web developers, and search engine optimization (SEO) specialists. As recently as the 1980s, the vast majority of people would have no inkling what that suggested, though they would suspect that spiders were somehow involved. It transformed the entertainment industry with dubious social media influencers taking on roles traditionally reserved for more conventional celebrities.

In transportation, vets, farriers, and groomers gave way to mechanics, insurance agents, and emissions testers. Engineering jobs in many fields have exploded.

So, it will be with AI. It will reshape many existing jobs, but it will also be a catalyst for entirely new ones. We have already seen the emergence of job postings searching for prompt engineers. These individuals can craft prompts that will guide large language models (LLMs) to deliver the desired output, from product catalogs to customer service answers, from aids to help to quit smoking to writing apps for mobile phones.

Artists are becoming prompt engineers, too, as they tweak and tune the output of AIs such as Midjourney until they get the image that they want.[7,8] It is a blend of creativity and technology, potentially a new form of artistry. The music industry will be next. Musicians might soon craft songs and symphonies by guiding AI with carefully chosen prompts.

Why stop there? The multi-modal approach we have started with combines text prompts with art and music. We can go further and explore the full synesthetic experience, crossing the boundaries between sight, sound, movement, or cognitive processes. Imagine a future where we can translate movements into music, turning a dance routine into a unique soundtrack, or an abstract painting that can be turned into an aural symphony? In this future world of AI, the possibilities are as vast as our imagination; who knows what new professions and creative avenues we might discover?

A critical emerging role will be that of an ethics director. Professionals will need to understand the ethical implications of AI systems ensuring fairness and accountability. In earlier chapters, we saw that a disciplined approach to the development of new technologies ensures growth that is more beneficial to society. This role is closely related to bioethics committees, which are composed of experts in medicine and science, and also to compliance and inspection offices in areas such as nuclear safety and environmental protection.

As AI systems become more complex, it is not too fantastic to think that one of the new professions of the next decade will be AI psychiatrists. Not that AIs will treat humans, but humans will help AIs evolve and become healthier. Scientists who develop AI systems today are aware of many ways AI can exhibit unexpected or undesired behavior. They may exhibit unintentional bias towards certain groups or react unexpectedly to novel stimuli. Imagine an autonomous car unable to decide how to handle a tortoise on the road or unable to decipher complex signage or situations. AIs will also make decisions that may appear unfathomable and may require a psychiatrist to delve into their inner workings to make it more understandable.

For instance, take an AI that's been trained to identify animals in rainforests but suddenly starts misclassifying them. The AI psychiatrist could use a diagnostic tool, much as a physician would use an MRI or an EEG, to map out what each of the layers in the network is doing, such as what kinds of input representations they are holding.

The psychiatrist could also engage in counseling, interacting with the AI, perhaps like we currently do with large language models like ChatGPT. Just as a human psychiatrist might use therapy sessions to guide a patient towards healthier thought patterns, an AI psychiatrist could use training data to "counsel" an AI. This could involve feeding the AI new data to help it learn from its mistakes or adjusting the data it's already learned from to correct biases or errors. For example, if an AI developed to screen job applications shows bias against certain groups. In that case, an AI psychiatrist might "counsel" the AI by providing it with a more diverse set of training data, helping it make fairer decisions.

The psychiatrist might sometimes come to the conclusion that surgery is needed. If the AI system's structure has to be modified, which could result in adjusting the weights in the neural network or pruning some connections, they could work with AI scientists in performing the operation.

It is often very tempting start a model from scratch. While that may be possible today, even with models like ChatGPT that show glimmers of emergent behavior, the cost of doing so for the complex models of the future may necessitate psychiatric intervention rather than a do-over.

Societal Change

With the spread of the industrial revolution, our ability to alter our planet changed significantly. No longer were we solely dependent on the strength of human and animal muscles to accomplish the feats of moving heavy objects, cultivating sustenance, and constructing shelters. Instead, harnessing steam power and subsequent electrification propelled us to new energy frontiers, elevating the collective standard of living for much of humanity. We can travel faster and farther than ever before, an exponential progression from traveling by foot to riding on horseback, from sailing ships to flying airplanes, and now venturing into space. We can also see both farther and more closely, from telescopes to corneal reshaping laser surgery, and from microscopes to MRIs.

Then, with the computer revolution, we transitioned from typewriters to computer screens and from ledger books to Excel. In manufacturing, robots are programmed to take on such physical tasks instead of hiring people to assemble cars on an assembly line, one person welding on the door, another attaching the gas tank, and so on. If there weren't enough people to do the job, that was okay because lack of human resources could be substituted by programmed machines. We yielded information processing supremacy to machines, and we watched on in amazement as machines surpassed us in skills we had honed for millennia.

Each new revelation of knowledge transforms humanity's estimation of itself, often leading to a humbler stance. We cede a bit more of the fiction of what makes us unique or human. During the European Renaissance, the realization that we were not the literal center of the universe shook our belief that we deserved to be the metaphorical center of the universe. Over the next few centuries, we learned that not only were we not the center of our solar system, but our solar system is an insignificant part of our galaxy, which is, in turn, an insignificant part of our local cluster, and so on, ad infinitum.

We still are unique beings who reason, though. We think logically, have leaps of intuition, and are the creators of the very things that surpass our physical abilities. We have the capacity for self-reflection, enabling us to question our place in the universe. Even the smartest, deterministically programmed computer could not create an architectural blueprint for a towering skyscraper, let alone orchestrate the logistics for making it a physical reality. The human spirit, it was believed, was needed to create music and reflect the world in art that would move hearts and minds. Computers could crunch data about voters' preferences, but political strategists had the skill to interpret the results and marry them with intuition to create election campaigns.

With AI, we are now embarking on a path that will take our creation beyond the essence of what allowed us to dominate the world—our intelligence. We are creating a technology in our image, one that may surpass our cognitive abilities. As we saw in the first chapter, AI is not merely a step on the road to automation. Even if computers are glorified calculators, AI goes several orders of complexity beyond that.

We already have AI systems that show signs of reasoning, not mere recall, regurgitation, or clever application of deterministic rules. Technological prodigies like GPT-4 are able to reason like a very smart person. Let's ignore for now the fact that they can also be very stupid and have no problem making stuff up. We are on the verge of artificial intelligence that is able to plan, strategize, and possibly generate works that would be considered genius if a human was the creator. When self-driving cars reach a critical count, AIs will plan traffic at city-wide and local levels. Trucks going cross-country will be driven and regulated by AIs. In schools, AIs will be able to monitor a child's progress in reading or math and optimize the pedagogical curriculum.

Could we be stepping into a technological utopia? It seems unlikely. History puts up a warning hand, reminding us how every major industrial revolution while heralding progress has also birthed new anxieties and novel misdemeanors. Stagecoach banditry gave way to high-speed car chases and haphazard gunfire exchanges. The information revolution inadvertently emphasized negative behaviors, as people found themselves ensnared by screens and engaging in rampant consumerism rather than being exclusively utilized for leisure. Freed time was often channeled toward extending work hours. However, on the brighter side, they did bask in the comforts of material abundance that would have been deemed royal luxuries in the past.

What does the evolution of genuine artificial intelligence reveal about us? What is human identity and self-worth? If we can build minds that are much better at reasoning, remembering, synthesizing, and imagining than us, then maybe we are not as special as we imagine ourselves. A profound societal convulsion could well be on the horizon as we grapple with this revelation, and such advancements in AI will likely face considerable resistance.

The crux of the issue lies in the way AI reframes the human condition. For centuries, we have been accustomed to the idea that our intelligence sets us apart from other life forms. We have always taken pride in our abilities to reason, learn, remember, and create. But what happens when we engineer minds that can outperform us in the very areas we've considered our unique domain?

But this doesn't have to be a crisis of identity. Instead, it can be an opportunity to expand and evolve our understanding of ourselves. The advancement of AI forces us to re-evaluate what we value in being human. It pushes us to move beyond intelligence as the primary measure of worth.

Final Thoughts

Whatever the future holds, we still have to work on our present problems. AI offers us a powerful tool to alleviate food and water insecurity for billions worldwide. It can minimize the toll of natural disasters by enabling early warnings and bolstering post-disaster resilience. It can enhance biodiversity in the sea, on land, and in the skies by enabling careful environmental monitoring and intelligent responses to changes. In tandem with other medical technologies, AI can refine patient care, enhancing life quality for many. As AI transforms education, employment, and other societal pillars, we have the power to shape these changes for the better. Ultimately, the positive impact of AI on society lies in our hands and depends on our collective will to harness its potential responsibly and ethically.

The next several decades will be tumultuous, with society evolving into new forms, and undoubtedly facing new crises. The powerful technology we are creating now, if handled with wisdom and foresight, may enable our children and their children to live in a world that will be more just, equitable, and harmonious than the ones their history books will talk about.

References

1. Kaku, M. *Quantum supremacy: How the quantum computer revolution will change everything*. (Doubleday, 2023).
2. Sutor, R. S. *Dancing with Qubits: How Quantum Computing Works and How It Can Change the World*. (Packt, 2019).

3. How Secure Is My Password? | Password Strength Checker. *Security.org*. At <https://www.security.org/how-secure-is-my-password/>
4. Church, G. M., Gao, Y. & Kosuri, S. Next generation digital information storage in DNA. *Science* **337**, 1628–1628 (2012).
5. Shipman, S. L., Nivala, J., Macklis, J. D. & Church, G. M. CRISPR–Cas encoding of a digital movie into the genomes of a population of living bacteria. *Nature* **547**, 345–349 (2017).
6. Service, R. DNA could store all of the world's data in one room. *Science* (2017). doi:10.1126/science.aal0852
7. Sheth, S. Top 5 Midjourney Artists using Artificial Intelligence to push the boundaries of creativity - Yanko Design. (2023). At<https://www.yankodesign.com/2023/05/22/top-5-midjourney-artists-using-artificial-intelligence-to-push-the-boundaries-of-creativity/>
8. O'Leary, L. A.I.-Generated Art Has Crossed the Uncanny Valley. *Slate* (2022). At <https://slate.com/technology/2022/09/ai-artists-colorado-art-competition-midjourney.html>

Index

O

P